Environmental Challenges and Solutions

Volume 2

Series editor

Robert J. Cabin, Brevard College, Brevard, NC, USA

Aims and Scope

The *Environmental Challenges and Solutions* series aims to improve our understanding of the Earth's most important environmental challenges, and how we might more effectively solve or at least mitigate these challenges. Books in this series focus on environmental challenges and solutions in particular geographic regions ranging from small to large spatial scales. These books provide multidisciplinary (technical, socioeconomic, political, etc.) analyses of their environmental challenges and the effectiveness of past and present efforts to address them. They conclude by offering holistic recommendations for more effectively solving these challenges now and into the future. All books are written in a concise and readable style, making them suitable for both specialists and non-specialists starting at first year graduate level.

Proposals for the book series can be sent to the Series Editor, Robert J. Cabin, at cabinrj@brevard.edu.

More information about this series at http://www.springer.com/series/11763

Stephen F. McCool • Keith Bosak
Editors

Reframing Sustainable Tourism

Editors
Stephen F. McCool
University of Montana
Missoula, MT, USA

Keith Bosak
University of Montana
Missoula, MT, USA

ISSN 2214-2827 ISSN 2214-2835 (electronic)
Environmental Challenges and Solutions
ISBN 978-94-017-7208-2 ISBN 978-94-017-7209-9 (eBook)
DOI 10.1007/978-94-017-7209-9

Library of Congress Control Number: 2015947818

Springer Dordrecht Heidelberg New York London

Printed on acid-free paper

Springer Science+Business Media B.V. Dordrecht is part of Springer Science+Business Media (www.springer.com)

Foreword

Tourism is a complicated, global, and growing phenomenon. The basic premise is that travel provides benefits, some to the traveler (who seeks personal benefits at a destination) and some to a destination host (who seeks personal benefits provided by the traveler). One of the substantial unifying concepts in the last three decades has been that of sustainable tourism. This book seeks to move beyond socially acceptable, ecologically responsible, and economically viable tourism ideas by asking fundamental questions. Case studies are used effectively to outline issues from many areas such as Brazil, Canada, India, Jamaica, Nepal, the USA, Vietnam, and several countries from Southern Africa.

One issue discussed is that of scale, such as geographic scales, but also timescales. This book makes a point that many sociopolitical systems do not always operate on linear relationships, but can be affected by sudden and unpredictable influences. Technological changes can have massive impacts on relationships and sharing of information, which quickly changes key aspects of the delivery of tourism benefits.

This book concentrates at the community scale, largely in the sense of the local community that serves as the host for the visitors. The book concentrates on the supply part of tourism, leaving tourism demand for another day.

An important issue is the definition of community. In this book, it tends to be geographical in concept, rather than professional or religious. But the scope of what is in or out of a community boundary is left open-ended, as maybe it must be if the concept is to be used widely.

The book makes it clear that tourism is just one of many human activities that affects host communities. Industrial, agricultural, and urban developments also occur in the same area and cause major changes. It is often difficult to disentangle what activity caused what impact. Also, positive impacts to one person or group may be linked to negative impacts to another person or group. Impacts seen as being positive by one person may be seen as being negative by another person.

There are examples given about the impacts caused by a lack of tourism, such as the cessation of tourism. These help to identify those complicated and often unmeasured impacts that occur when a constant flow of visitors move through a community and are largely taken for granted. One of the most striking examples of this, known to his observer, was the closure of most rural tourism in England and Scotland in 2001 when the hoof and mouth disease hit domestic livestock. This was an unintentional experiment that revealed the massive positive impact of rural tourism and the general lack of understanding of its importance, once it was gone. It was revealed that tourism in rural Scotland was worth 5 % of gross domestic product, while agriculture was worth much less at 1.4 %. So government actions of stopping rural tourism in order to help agriculture damaged the much more economically valuable activity (Stewart 2002). One result on UK government policy was changes in agricultural subsidies, moving benefits from agricultural food production to landscape quality enhancement.

I once talked to a man picketing at the opening of the new visitor center in Algonquin National Park in 1993. He was a part-time park staff member complaining about the reduction of jobs in the park due to government funding reductions. He stated that there are only two types of jobs in his area, tourism in the summer and logging in the winter. This statement puts into focus the complexity of human activities in this area. One activity, logging, is portrayed as sustainable by the foresters but portrayed as unsustainable by park visitors. The other activity, tourism, is seen as being sustainable by park visitors, but damaging by the loggers due to the continual pressure by park visitors to stop logging in the park. So understanding the financial sustainability of this man was complex and fraught with definitional issues. Sustainable in what way, of what resource, and to whom are basic questions.

A major issue is who benefits and who pays. This of course is an issue with all economic activities, whether it is the development of a mine, a logging activity, or tourism. Much tourism analysis, but not in this book, ignores that tourism is just one of many other competing economic activities in an area. In fact, tourism may be the most sustainable activity over the long term, in that the benefits can continue indefinitely if properly managed. This cannot be said for mining and some manufacturing industries; these come and go.

Some aspects of sustainability can be bipolar, providing both benefits and disbenefits at the same time. For example, some aspects of culture are changed due to visitors' ideas, but visitor demands for observation of local culture provide incentives for preservation of some aspects of that culture. Such inherent dichotomies are rife in the application of sustainable tourism concepts.

Sustainable tourism, a laudable goal, is essentially socialistic; it espouses community management and control, where individual selfish desires are balanced against community desires and benefits. Many of the papers in this book look toward methods for such control, whereby plans are made and implemented. This only occurs where there is the power and institutional structure to implement those plans.

One of the methods for control is the use of standards and certification of various elements of the tourism supply chain, such as for transportation operations, tours operators, accommodation providers, and food suppliers. These activities attempt to ensure that worthwhile overarching standards are implemented. There is evidence that travelers expect base levels of standards, so that they are safe from the dangers of transportation, accommodation, and food hazards while on vacation. They may not be supportive of financial sustainability of local communities if the consumer sees low prices as a primary goal.

I wonder if sustainability is not an end, but only a process. One can never fully achieve the elusive goal. It is always like the mysterious blonde woman in the white thunderbird in the 1973 movie *American Graffiti*. She is always a dream, out there for sure, but one can never get as close as wanted.

Paul F.J. Eagles

Reference

Stewart, W. 2002. *Inquiry into foot and mouth disease in Scotland*. Edinburgh: Royal Society of Edinburgh. http://www.scribd.com/doc/61435188/Foot-and-Mouth-Disease-in-Scotland. Accessed 27 Jan 2015.

Preface

If we have learned anything about the search for sustainable tourism, it is the importance of asking the right questions. In this text, we agree with American songwriter and singer Bob Dylan that "The Times, They are a-Changin" written in response to the political and social contentiousness of the 1960s and yet remarkably applicable to the early twenty-first century. We believe the times are still "a-Changin," and therefore we ask here if the conventional paradigm of sustainable tourism, that which occurs at the intersection of what is socially acceptable, ecologically responsible, and economically viable, is really an appropriate and effective way of conceptualizing tourism development in the twenty-first century? We acknowledge this paradigm, during the first quarter century of discussion about the objects of economic development, which focused a great deal of scientific, philosophic, and development activity on reducing the negative impacts of tourism.

But from those experiments, we have learned much, so much that we wonder if a more contemporary approach to tourism would be more effective as well as more in line with our knowledge of tourism as a component of a social-ecological system as well as the incredible complexity of the global development system. Complexity and uncertainty greet us at every social-organizational, temporal, and spatial scale. What is economically viable at one scale is not at another. What is socially acceptable to one group is not for another. And what might be viewed as ecologically responsible may be seen as negligent by others.

Given the pace, scale, and type of change we see today, can our efforts benefit from alternative conceptualizations of sustainable tourism? We believe so. This book is about thinking a bit differently about the aims and tools for sustainable tourism. It is written principally for a North American audience because we see a lack of attention to sustainable tourism in that geography. That said, the text uses a number of examples from other continents to illustrate new perspectives, some critical, others more helpful.

The book is organized into four sections. The first provides some foundational material about sustainable tourism, how the concept developed historically and how our thinking about it has evolved. In the second section, several authors provide some frameworks about different arenas of sustainable tourism that can help us think differently and be more effective in achieving the goals of sustainable tourism. The third section contains several case studies reflecting ways of thinking differently and the implications of doing so. In the final section, we suggest a way forward that we believe will more effectively help tourism development build communities, advance opportunities for higher-quality visitor experiences, and protect our natural and cultural heritage.

We want to thank each of the contributors to this volume, because each not only spent time writing up their stories but also have furthered different notions of sustainable tourism. We also want to thank our editors and publishers at Springer who helped bring this idea to a fruition.

Missoula, MT, USA

Stephen F. McCool
Keith Bosak

Contents

Contributors

Valéria de Meira Albach Departamento de Turismo, Ponta Grossa State University, Ponta Grossa, PR, Brazil

Rosemary Black School of Environmental Sciences, Charles Sturt University, Albury, NSW, Australia

Keith Bosak University of Montana, Missoula, MT, USA

Kelly Bricker Department of Parks, Recreation, and Tourism, University of Utah, Salt Lake City, UT, USA

Robert C. Burns School of Forestry and Natural Resources, West Virginia University, Morgantown, WV, USA

Jason A. Douglas Department of Environmental Studies, San Jose State University, San Jose, CA, USA

Rajinder S. Jutla Missouri State University, Springfield, MO, USA

Laura Lawton Griffith University, Gold Coast, Australia

Tuan-Anh Le Ha Noi University, Hanoi, Vietnam

Stephen F. McCool University of Montana, Missoula, MT, USA

Barbara Jean McNicol Mount Royal University, Calgary, Alberta, Canada

Jasmine Cardozo Moreira Departamento de Turismo, Ponta Grossa State University, PR, Brazil

Art Pedersen Independent Consultant, Helsinki, Finland

Martina Shakya Institute of Development Research and Development Policy, Ruhr University Bochum, Bochum, Germany

Susan Snyman Environmental Policy Research Unit, University of Cape Town and Wilderness Safaris, South Africa

David Weaver Griffith University, Gold Coast, Australia

About the Authors

Valéria de Meira Albach Ms Albach is assistant professor of the Tourism Department at Ponta Grossa State University, in Paraná, Brazil. She has a master's degree in geography and specialization in environmental geography and holds a PhD from the Federal University of Paraná. She works as a tourism consultant for protected areas, including the Superagui National Park, also in Paraná, Brazil.

Rosemary Black Dr Black gained her PhD from Monash University where she undertook research on ecotourism. She currently serves as associate professor in the School of Environmental Sciences, Charles Sturt University, Australia. Rosemary's teaching focuses on tour guiding, environmental interpretation, adventure tourism, and outdoor recreation. She conducts research in tour guiding, interpretation, ecotourism, and social dimensions of natural resource management. Most is applied research that she undertakes in partnership with industry associations, protected area managers, and communities. Prior to being an academic Rosemary worked in the fields of adventure tourism, protected area management, and recreation planning. She serves on the boards of Guiding Organisations Australia and Interpretation Australia.

Keith Bosak Dr Bosak earned his PhD in geography in 2006 from the University of Georgia where he studied the local effects of global conservation policy in the Nanda Devi Biosphere Reserve. As a geographer, he currently serves as associate professor at the Department of Society and Conservation at the University of Montana. Dr Bosak's research interests are broadly centered on the intersection of conservation and development, and as such, he often studies nature-based tourism and sustainable tourism in the context of development and protected areas. He has conducted research on ecotourism and environmental justice in India, scientific tourism in Chile, and geotourism in Montana. Aside from tourism, Dr Bosak has conducted research on climate change impacts and adaptations among tribal populations in the Himalaya, private protected areas in Chile, and conservation and development initiatives in Montana. Since 2005, Dr Bosak has been instructing field courses on tourism and sustainability in the Garhwal Himalaya. Dr Bosak and his wife Laura founded the Nature-Link Institute, a 501(c)3 that seeks to reconnect people with their environment through research, education, and advocacy.

Kelly S. Bricker Dr Bricker serves as professor and interim chair of the Department of Parks, Recreation and Tourism, University of Utah, USA. She completed her PhD research with the Pennsylvania State University, specializing in nature-based tourism. She has research/teaching interests in ecotourism, sense of place, resource management, and environmental and social impacts of tourism. She has conducted research on certification, tourism and quality of life, heritage tourism, social impacts of tourism, and impacts on natural resource tourism environments. Since 1982, Kelly has worked in ecotourism and adventure programs such as the Florida High Adventure Sea Base, Sobek Expeditions, World Heritage Travel Group, and Rivers Fiji. She serves on the boards of The International Ecotourism Society and Global Sustainable Tourism Council.

Robert Burns Dr Burns is associate professor at West Virginia University with nearly two decades of research experience in public land social science and in working with public land managers. Dr Burns has conducted research focused on all aspects of monitoring visitor use in federal and state land and water-based settings. He is responsible for an international research program in Brazil, focusing on visitor monitoring in parks and protected areas. His expertise is in quantifying motivations, benefits, use patterns, behaviors, and other social carrying capacity indicators in outdoor settings.

Jason A. Douglas Dr Douglas is an environmental psychologist who conducts research with underserved communities to develop an understanding of environmental issues at the local, state, federal, and international levels. He strives to use research to educate and empower individuals exposed to social, economic, and environmental disparities. Dr Douglas has over 8 years of experience in community-based participatory research and evaluation. He has worked on a wide range of projects concerning the political ecology of participatory conservation in Jamaica, environmental justice and education in underserved urban communities, public policy and advocacy efforts to address root causes of childhood obesity, and hospital system improvements related to homeless populations.

Rajinder S. Jutla Dr Rajinder S. Jutla is a professor and director of the Planning Program in the Department of Geography, Geology, and Planning at Missouri State University, Springfield, Missouri, USA. His undergraduate and graduate studies are in the areas of architecture, landscape architecture, and planning. His research interests include history and theory of urban design, historic preservation, environmental perception, urban tourism, and cultural geography. He has presented papers at a number of national and international conferences and has published journal articles and book chapters in these areas.

Laura Lawton Laura Lawton is an associate professor within the Department of Tourism, Sport and Hotel Management at Griffith University, Australia. She has published numerous government reports, academic journal articles, and book chapters in several areas, including protected areas, ecotourism, resident perceptions of tourism, and cruise ship tourism. She is a coauthor of the tourism text *Tourism Management*, published by Wiley, and sits on the editorial boards of four academic journals. Laura serves as deputy chair of The International Centre of Excellence in Tourism and Hospitality Management Education (THE-ICE), an independent international accreditation body that specializes in tourism, hospitality, and events education.

Tuan-Anh Le Dr Tuan-Anh Le holds a PhD from Griffith University, Australia. Currently, he works as a lecturer at the Faculty of Management and Tourism, Ha Noi University, Viet Nam. His area of research interests covers community-based tourism and sustainable development. Over the past 15 years, Mr. Le has been active in the provision of training and consulting services for tourism projects in Viet Nam funded by international organizations, for instance, the EuropeAid, Asian Development Bank (ADB), UNESCO, the Netherlands Development Organization (SNV).

Stephen F. McCool Steve is professor emeritus at the Department of Society and Natural Resources at the University of Montana in Missoula, Montana, USA. He has been active in research and development activities in the area of tourism and visitor management in protected areas for over 45 years, contributing frequently to a variety of journals and conferences. In addition, he has worked to change how managers approach public engagement and protected area planning processes, frequently arguing that we need to dive deeper and think differently in a world of change, complexity, and uncertainty.

Barbara Jean McNicol Dr Barbara McNicol is an associate professor of geography, past 6-year chair of the Department of Earth Sciences, and assistant director of the Institute for Environmental Sustainability (IES) at Mount Royal University in Calgary, Alberta, Canada. She is also an adjunct associate professor of geography at the University of Calgary where she has taught in both of the international relations and geography degree programs. Barbara is a social environmental geographer conducting behavioral research at the interface between sustainable tourism planning and management and environmental and natural resource management while emphasizing parks, recreation, and community land use. Barbara has worked with the local government, such as the Town of Canmore, Economic Planning and Development Office, in Alberta, and with Parks Canada Agency in the national parks of Pacific Rim, Banff, and Jasper as well as independently as a tourism and environmental research consultant.

Jasmine Cardozo Moreira Jasmine Cardozo Moreira is associate professor and chair of the Tourism Department at Ponta Grossa State University, in Brazil. Her expertise is in community development, focusing on human dimensions of tourism planning in Brazil National Parks and geoparks worldwide. She received a bachelor's in tourism, PhD in geography, and a postdoctoral study in Spain. She is an active member of the federal academic accreditation team that evaluates tourism programs throughout Brazil. Also a member of the International Academy for the Development of Tourism Research in Brazil, she provides academic consulting to the US Forest Service International Programs, the Brazil Park Service (ICMBio), and the Erasmus Program.

Arthur Pedersen Art Pedersen is a consultant on protected areas and tourism development located in Helsinki, Finland. His professional experience includes work on tourism, protected areas, and economic development issues. He has been a program specialist with UNESCO's World Heritage Centre (WHC) in Paris. Other professional experience includes work in Latin America, Europe, Asia and Africa, and the Middle East with regional tourism strategies, national park management plans, tourism assessment and feasibility studies, creation of regional tourism organizations, and the development of practical tourism marketing-promotional strategies. Many of these activities involved the push and pull of economic development and environmental and cultural impacts. He has worked on issues of visitor limits in fragile areas and in linking tourism benefits to local communities with the goal to aid protection and conservation efforts.

Martina Shakya Dr Martina Shakya is senior research fellow at the Institute of Development Research and Development Policy at Ruhr University Bochum (RUB), Germany. She holds a master's degree in geography, anthropology, and political science from Münster University (1996), received postgraduate training in international development at the German Development Institute (GDI, 1998), and a doctorate in human geography (2009) from RUB. She has extensive practical work and research experience in various parts of the developing world and worked with government and nongovernment organizations in Nepal and South Africa as an advisor for sustainable tourism development.

Susan Snyman Susan Snyman holds a PhD (economics) from the University of Cape Town (UCT). Having completed PhD coursework at the University of Goteborg in Sweden, the focus of her PhD research was on the socioeconomic impact of high-end ecotourism in remote, rural communities adjacent to protected areas, based on over 1800 community interviews in six southern African countries. Sue has 15 years of experience in the ecotourism industry in southern Africa and is now the Group Community Development and Culture manager for Wilderness Safaris, as well as the regional director of Children in the Wilderness. Other positions include vice-chair of the IUCN WCPA Tourism and Protected Areas Specialist Groups and a research fellow at the Environmental Policy Research Unit at UCT.

David Weaver David Weaver is professor of tourism research at Griffith University, Australia, and has published more than 120 journal articles, book chapters, and books. He maintains an active research agenda in sustainable destination management, ecotourism, and resident perceptions of tourism. Professor Weaver has contributed extensively to leading journals, and his widely adopted textbooks include *Ecotourism* (Wiley Australia), *Encyclopedia of Ecotourism* (CABI), and *Sustainable Tourism: Theory and Practice* (Taylor & Francis). He is a fellow of the International Academy for the Study of Tourism and has delivered numerous invited keynote addresses around the world on innovative tourism topics.

Part I
Foundations

Chapter 1
Sustainable Tourism in an Emerging World of Complexity and Turbulence

Stephen F. McCool

Abstract The world is changing and so too are our notions of sustainability, sustainable development and sustainable tourism. After pointing out how a small Montana community has changed over the last 125 years or so, I raise fundamental questions about what it means to be sustainable, and how limited conventional definitions of sustainable tourism are in the complex, fast changing world of the twenty-first century. And, as I note, these questions also face small rural communities in developing regions of the world. Rising complexity leads to accelerating uncertainty requiring us to rethink the mental models we use to survive in the world, and that is what this book is about—reframing conventional notions of sustainable tourism to something more useful and appropriate.

Keywords Systems thinking • Whitefish • Resilience • Complexity • Mental models

1.1 Whitefish, Montana: A Story of Sustainability?

In the late nineteenth century in northwestern Montana, a new town arose in what was then what we would call now wilderness. The town was situated in a broad and forested valley, containing lakes and numerous streams and rivers. To the east rose the Rocky Mountains and the Continental Divide. To the west, travelers would find more forests and streams, but deep gorges and spectacular mountains intervened before they gained the gentler and more open setting of the Columbia Basin.

The town was founded initially on fur trading and then on its access to timber resources. Sawmills popped up there to process the timber needed to build farms and homes and industry in western Montana. Shortly after, the Great Northern Railway was built, and because of its strategic location, expanse of flat land and other features, the town became a division point for the railroad, creating jobs and

S.F. McCool (✉)
University of Montana, Missoula, MT, USA
e-mail: Steve.McCool@cfc.umt.edu

S.F. McCool, K. Bosak (eds.), *Reframing Sustainable Tourism*,
Environmental Challenges and Solutions 2, DOI 10.1007/978-94-017-7209-9_1

employment for the new community. For many decades, because of the seemingly inexhaustible supply of timber and demand for goods to be transported, the railroad and lumber provided the economic foundation for this community. For several years, it was known, appropriately, as Stumptown, and then, in the early twentieth century, its name was changed to Whitefish. While railroad, logging and mill work could be dangerous occupations, they paid good wages and the industry attracted many workers and their families to the Whitefish area and it prospered. Although the population of the town waxed and waned over the years in response to the regional and national economic situation, the diverse economic base enabled it to "roll" with these changes. In the face of change, Whitefish persisted and avoided the decline that characterized many of its sister communities in the Rocky Mountains.

In the late 1940s, the town's economic base began to diversify. Several visionary individuals started a ski resort just north of town. For several decades, the ski area grew, but slowly, periodically adding ski runs, lodges, accommodations and associated amenities. And in the 1960s, a large scale aluminum processing plant was built just to the east of town in the nearby community of Columbia Heights. Large amounts of power, generated by the Hungry Horse dam which flooded the south fork of the Flathead River valley about 20 miles to the east, provided low cost energy. The plant furthered the manufacturing dependency of the community.

In the 1970s, as timber harvest volumes increased on the national forests surrounding Whitefish, criticisms and legal challenges to the Forest Service management regime increased. Expressions of public concern grew in scope and intensity; recreation and wilderness, esthetic quality, habitat for endangered species such as the grizzly bear and a host of other values were increasingly identified as key outputs of national forests, outputs that were often threatened by high levels of timber harvesting. The allowable cut on the surrounding Flathead National Forest dropped dramatically in the mid 1980s, and with this came a decline in the predictability of timber supplies, which had served as an underlying assumption of the Whitefish and Flathead Valley economy. Mills increasingly competed with each other, and for several years, whole logs were shipped to Japan for processing, and then returned to the U.S. as plywood. In a sense, this was a fortunate, for in the mid 1980s, the timber industry began to decline. Harvestable volumes on the surrounding Flathead National Forest were initially limited, and then experienced declines to less than 10 % of its historical average.

In the 1980s, however, as the Canadian dollar dropped in value and as timber processing became increasingly tenuous, the ski area grew dramatically, and as other destination areas in the US, it expanded to include condos, trophy homes and more opportunities for employment. To many, the ski resort represented a dramatic change, not only in source of employment and income, but quality of life, culture and tradition. The loss of resource-based jobs and the corresponding increase in service jobs, primarily in the tourism sector was a radical transformation for the community. The ski-area presaged golf courses, resorts, gated communities, celebrities and tall, skinny lattes. RVs had replaced logging trucks, cowboy hats over corked boots, and skis rather than chain saws.

Over a century after the forests were cleared for the townsite, Whitefish still exists. But its history and evolution is the story of hundreds of small towns and villages in the Pacific Northwest, if not in detail, then in general outline: development of resource commodity based industries—agriculture, range, timber and mining—a long period of dependency on those industries, followed by a decline—sometimes gradual, sometimes precipitous. Many communities, like Whitefish, survived this period through fundamental restructuring of their economic base, primarily by developing a vibrant tourism industry. Many other communities did not survive, and exist only in the memories of their former residents and in photos located in state historical society archives. And still many others remain in a languid, stagnant condition, generally characterized by low incomes, high average age of residents, lack of opportunity for children and significant outmigration.

While the architectural theme of contemporary downtown Whitefish reflects its early history and timber influence, there is a question of whether Whitefish can be depicted as a sustainable community. Given the 100 plus years of its existence, how could one really question whether the community has been sustained? But what does it mean, conceptually and practically, to be a sustainable community? Given the importance of recreation in the recent history of Whitefish, can that development be termed "sustainable tourism"? If so, what has tourism sustained? Have residents seen incomes grow, the environment preserved, the community become more vibrant, opportunities enhanced? Are nearby natural sites, such as Glacier National Park, the Bob Marshall Wilderness, even the Flathead National Forest, more capably and sustainably managed? Is Whitefish a safer place to live, work, and raise one's children? What role has tourism played in these changes? Could we make tourism more effective in dealing with crime, unemployment and access to education and health care? If so, how? Is there even a connection? And what about the future—what role would tourism play in the face of climate changes that might impact snow skiing opportunities, cause the glaciers in Glacier National Park to disappear, or impact the region's outstanding cold water fisheries—all being the foundation for Whitefish's twenty-first century tourism industry.

The story of Whitefish, is then very revealing in that it illustrates the changes in relationships between society and the environment. As our awareness, sensitivity, and knowledge of society-environment links grow, we confront fundamental questions about what futures face us and our children. Does the history of Whitefish reveal a community that has sustained itself over the last century? Are choices broader or more constrained than in the past? What would it mean to answer in the affirmative or the negative? Given the dramatic nature of social and economic change that will inevitably come, will Whitefish be sustainable in the future?

In thinking about such meta questions, we confront numerous corollary questions; e.g., what is and/or should be the relationship of Whitefish to external economic forces? How might Whitefish best buffer itself from the inevitable, but unknown events of the future to ensure its sustainability? How adequately have issues such as inequities in income distribution, access to public decision-making, or environmental impacts been resolved; how should they be addressed in the future? How have decisions taken—locally or nationally, intentional or unintentional—affected quality of life and for whom? Who has had access to, and influence upon, these decisions? These are difficult, but unavoidable questions; if we fail to address them explicitly, we risk negative outcomes.

The purpose in illustrating Whitefish has been to indicate both the complexity and ambiguity of the concept of sustainability, and as we shall see later, sustainable development. While we can view sustainable tourism as a kind of marketing and promotional device, playing on genuine concerns for environmental and social impacts, can the concept be reframed in ways useful for twenty-first century challenges and opportunities?

These questions are as important for the people of Maguary village situated alongside the Tapajos River in the lower Amazon basin of Brazil as they are for residents of Whitefish. The ribeirinhos (river people) of Maguary are surrounded by the Tapajos National Forest and bounded by the river. They are isolated and buffered by these outstanding natural resources, for just outside the National Forest, conversion of Amazonian rain forest to soybean fields is nearly complete, generating a whole new combination of industry and business. And while this new economic development may provide jobs and revenue for the small village, they may also threaten its quality of life. As one villager told me "the national forest protects our daughters and village" from the negative consequences of development. And yet, the ribeirinhos see sensitive, small scale tourism as a livelihood to keep their village the way they want it. For them, sustainable tourism is more than livelihood though, it helps buffer their way of life from modern development.

1.2 A Growing Sense of Unease

Whitefish and Maguary are as far apart socially, politically, environmentally and economically as two communities can be, and yet they bear the same concerns and are confronted with a similar sense of unease about their futures. In recent years, there has been an increasing suspicion that the actions of society—laws, technology, politics, human behavior, conflict, consumption, development—lead to unprecedented levels of turbulence, chaos and disorder. Not only do everyday people carry anxieties that in the headlong rush to adopt new technology, develop an industrial base or capture the money wealthy tourists are all too willing to part with, that important values, qualities and resources are being irretrievably lost, but there seems to be accelerating apprehensions we may be irretrievably impacting resources and values. Are we eroding the very capital upon which society is founded—ecological processes, structures and components—that if our behavior is continued will deprive future generations of any positive legacy?

In the face of such difficult and troubling questions, many have turned to the concept of sustainability as the light to show a pathway to the future. At the core of this concept is the belief that the actions of present-day societies adversely and unacceptably alter the fundamental life support systems upon which cultures depend and the moral norm that the present generation should not adversely affect the choices and options of future generations. Believing both in the unacceptably of human-induced impacts and the morality of leaving choices to our children and grandchildren fuel the drive toward sustainability and as a result, the sustainable tourism.

1.3 Why Study Sustainability and Sustainable Tourism?

There is great concern that the legacy our societies are leaving future generations is not one they would enjoy, or even the one we have received from our predecessors, that choices are more limited, that problems and challenges constrain what they will be able to do, and that opportunities for improving one's lot are more constrained than ever. To sustain something means that we provide the future with at least a range of choice equal to or more than what we have experienced. Increasing biophysical impacts of human activity such as climate change and loss of biological diversity may lead to fewer options and less opportunity. Thinking in terms of sustainability many people propose will help stem these losses and leave a legacy of choice and environmental quality to the future.

However, sustainability, sustainable development and sustainable tourism are concepts that while perhaps easily understood, at least in the underlying motivation, are much more challenging to apply. What is sustainable to whom for what reason may vary from group to group, person to person, society to society. It is not a matter of simply practicing conservation as Butler (2013) argued. What is conservation and what is sustainability are socially complex questions that defy simple minded answers inappropriate for the turbulence of the twenty-first century.

Underlying notions of sustainability are important human values such as equity, trust, ownership, ethics, empathy and so on. Sustainability is as much about our values as it is about the technology and expertise needed to apply those values. To practice sustainability that is real and not an illusion means we need a better understanding of what it is we seek and more critical examination of what it is we mean. And, given that concepts vary over time and space, change in relation to new information, and evolve as we gain wisdom, there is always the question of appropriateness of current paradigms.

So, we need to ask ourselves questions about sustainability, sustainable development and sustainable tourism. We need to know what these concepts mean, what challenges exist in their application, and how they might be changed in response to changing conditions and social preferences. That is what this book is about.

1.4 Emerging Challenges in a World of Complexity, Change and Uncertainty

The world of the twenty-first century is significantly different from the world in which the concept of sustainable tourism originally emerged. First, it is much more globalized, with connections and interactions occurring over time and space at a pace that is difficult to understand and appreciate. Rapidly changing technology has shifted how people interact, how frequently and how much in depth. For example, Twitter has become a 140 character blitz of social change—even of rebellion—but vulnerable to hacking, misinformation and rumor. The world wide web puts information at people's finger tips, meaning that a small dude ranch in the Rocky Mountains of Montana competes with a game lodge in South Africa for clients. Second, rising interest in indigenosity leads to smaller units of governance—while protecting cultural heritage—but increasing transaction costs. Terrorism was of limited significance in the 1980s, but now has risen to a major violent force affecting safety and security across the globe. Our knowledge about how things work has grown dramatically with the rise of systems thinking popularized by authors such as Peter Senge, John Sterman, and Donella Meadows. This provides us with alternative frameworks to consider how we can enhance the human condition.

We recognize, more so than in the early 1990s that the implicit assumptions we hold about the world influence our behavior. The rise of systems thinking[1] (see Senge 1990 for example) together with the recognition that the mental models we carry around influence our behavior (and even the evidence we may see in scientific exploration) and the increasingly acrimonious debate about tourism development suggest that a critical examination of sustainable tourism concept is in order. Mental models are our simplified representations of reality that help us work through the complexities of not only everyday living but also the grueling problems of development, poverty alleviation and environmental protection. Mental models are frequently influenced by our successes of the past, and so strongly held they serve as barriers to seeing evidence that challenges those representations.

Conventional sustainable tourism mental models of the late twentieth century were constructed out of modernist and postmodernist assumptions that the world is predictable, linear, ultimately understandable and basically stable (see Kohl and McCool 2016 for a comprehensive discussion of these assumptions and their impact on planning for tourism in protected areas). This view of the world resulted in complex problems being reduced to "digestable" parts, with "solutions" to each component developed. After solutions for each problem component were solved, then components were put back together for a more "comprehensive" solution. These solutions often become the panaceas Ostrom (2007) critiqued in her insightful essay "Sustainable Social-Ecological Systems: An Impossibility?"

This reductionism produced policies and development activity that focus on interventions in communities in one particular sector—tourism in this case—with little understanding of the broader scale consequences, both positive and negative, resulting. For example, tourism interventions have been criticized as insensitive to indigenous community cultural norms and values, in other cases as producing low quality jobs, and in still others leading to unacceptable environmental impacts. Communities—even small ones—are incredibly diverse, in terms of norms, affluence, political power, access to education and health care, type of job and so on. Focusing on tourism as an intervention without a broad understanding of the entire system will likely lead to some stresses and strains that one could argue negate the benefit of enhanced income for some. In this sense, the solution, sustainable tourism,

[1]Loosely defined here as the study of how parts of the whole influence each other.

becomes the problem. We must be careful, in a more general sense that the solutions we implement today do not become the problems we must solve in the future.

Further, it is unclear how one would measure and assess whether a sustainable tourism initiative was successful and why: interventions often display confusion between inputs and outcomes, and the spatial, temporal and social-organizational scales are often unstated at which interventions are aimed often go unstated as well as not being subject to monitoring. Implementing an intervention might look good for a government program or to an NGO's donors, but how do we know it worked? For whom did it work? Who benefited? Who did not? Why? Further, a focus on sustainable tourism as small scale businesses or community tourism initiatives ignores both the idea of reducing the negative consequences of all tourism in general and how tourism development integrates into the larger economy of a village or region. In one sense, the goal of sustainable tourism has been to ensure economic stability, particularly at the community level—a goal difficult to achieve in a world of globalized financial institutions and processes.

Overly simplistic models and panaceas—such as finding the intersection of ecological sensitivity, economic feasibility and cultural acceptability—are deceptive. It is unlikely that economic, environmental and social acceptability concerns will be valued equally in sustainability discussions by different groups. How do constituencies differ in their preferences? Why? What about constituencies not yet alive—those generations the Brundtland Commission speaks to? Economic feasibility is so dependent on short-term market and financial conditions as to be counter to the long-term notion of sustainability as intergenerational equity. Social acceptability varies significantly across cultures and even within small communities, so we are confronted with the question of: acceptable for whom? There are no clear, technically based guidelines for answering this question. Dryzek (1987) argues in this context that social choices must be first ecologically rational, for if we lose the environmental basis for human life, there is no future for other considerations. One could argue that the illusion of achieving sustainability (through excessively reductionist approaches) may be its most fundamental obstacle.

We have now begun to realize that a new set of assumptions about the world would advance our opportunities to learn and produce insights more beneficial to tourism decision-making. First, we understand that the world is dynamically complex, that is, the world changes in a non-linear rather than incremental manner, that small changes in one variable may lead to large changes in another. Second, for all practical purposes, the world is impossible to completely understand, that is, there will never be enough data or science to completely explain the causes and consequences of events, patterns and structures. Third, the world is ever-changing, by this we mean that we can always expect surprises, that because knowledge is tentative and incomplete, unpredicted consequences will likely arise in places and at times we are least likely to expect. Finally, the world is connected as a giant complex adaptive social-ecological system, that numerous drivers and forces acting at the global level influence the effectiveness, usefulness and appropriateness of economic development actions at the local level. Such drivers include changing models of governance, population dynamics (e.g., growth, aging, migration), technological advancement, economic restructuring, climate change, and so on (see Kohl and McCool for more in-depth discussion on these assumptions).

Such systems are comprised of various actors, actions, resources, relationships and influences (Andereis and others 2004)[2]; they are subject to a variety of internally and exogenously induced perturbations; and they contain properties characterizing the whole in addition to their constituent parts. Relationships between causes and consequences are often mediated by a number of linkages, which means that cause-effect relationships are loosely rather than tightly coupled. Temporal delays between actions and effects may be long: actions taken by an entity may lead to effects thousands of kilometers distant. Interactions among different scales are typical. Problems apparently "solved" in one location may be simply, and sometimes carelessly, displaced to some other place, sometimes with less capacity to address those problems.

A systems lens holds a number of consequences, most notably an improved understanding of the notion of uncertainty; that the future is no longer a linear progression of the past; that use of systems thinking, particularly conceptualizing tourism as one component of a social ecological system, helps managers, practitioners and academics think more holistically about tourism; and that interests, while they vie and compete for attention and resources, may be better off building linkages, partnerships and relational capital (Nkhata et al. 2008) to secure progress in the face of contention, complexity, uncertainty and change.

Complex social-ecological systems are also characterized as containing emergent properties—attributes of the whole which cannot be predicted from understanding the parts. Just as an understanding of the biology of brain cells does not predict personality of a person, developing new and small tourism businesses may not predict the sustainability of the system as a whole or its ability to adapt to the inherent turbulence of twenty-first century earth. In this sense, assumptions about the homogeneity of small villages implicit in many development initiatives are often exposed by rancor, jealousies and conflict introduced by the perceived inequities in resulting incomes and opportunities.

These and many other changes mean that the world is turbulent and uncertain, with both challenges and opportunities abundant, but often difficult to understand. At the time the concept of sustainable tourism evolved, such turbulence and uncertainty was generally not recognized, and so sustainable tourism was conceived of something that was relatively simple to implement: just find the convergence of the socially acceptable, ecologically viable and economically feasible and ta da, sustainable tourism! We now know that such a conceptualization is overly simplistic, provides an illusion that finding this intersection can be done easily and hides important questions that are mainly value laden. The twenty-first century requires new ways of thinking to make progress, drive sustainable tourism toward more

[2] A social-ecological system is defined by Andereis et al. as "an ecological system intricately linked with and affected by one or more social systems."

useful conceptualizations and builds our abilities to think critically about social goals, societies' preferences, and social-ecological viability.

It is within this context that the concept of this book was born. We believe we need to reframe what sustainable tourism is about, its role in economic development and how communities can become more resilient and vibrant in an environment of turbulence, change and new ways of thinking. In short, a new mental model of sustainable tourism is required in a century typified by change, uncertainty and conflict.

The focus in this book is at the community scale and larger, and not at an individual business or individual scale although we do mention sustainability at those smaller scales. Codes of ethics and a good sense of social responsibility typify smaller scales providing a basis for individual and firm action. At larger scales, however, the complexities of human society and its interactions with the natural world remain a significant challenge demanding continuing dialogue, experimentation and adaptation. This is the focus of the book. It is organized into three sections: The first section emphasizes understanding the historical roots of the sustainable tourism concept, how tourism interacts with other domains of social activity and how it can be conceived of as an economic development tool. In the second section, invited authors propose frameworks to structure our thinking about how sustainable tourism can be realized in several important domains—protected areas that often form the basis of tourism attractions, communities, and the private sector. The third section of the book contains several diverse case studies of tourism development challenges and opportunities, each chapter concluding with some tools found useful in advancing sustainable tourism. We conclude the book by synthesizing important concepts and tools.

References

Anderies, J.M., M.A. Janssen, and E. Ostrom. 2004. A framework to analyze the robustness of social-ecological systems from an institutional perspective. *Ecology and Society* 9(1): 18. http://www.ecologyandsociety.org/vol9/iss1/art18/

Butler, R. 2013. Sustainable tourism—The undefinable and unachievable pursued by the unrealistic? *Tourism Recreation Research* 38(2): 21–26.

Dryzek, J.S. 1987. *Rational ecology: Environment and political economy*. Oxford: Blackwell.

Kohl, J., and S.F. McCool. 2016. *The future has other plans*. Denver: Fulcrum Press.

Nkhata, A.B., C.M. Breen, and W.A. Freimund. 2008. Resilient social relationships and collaboration in the management of social–ecological systems. *Ecology and Society* 13(1): 2. http://www.ecologyandsociety.org/vol13/iss1/art2/

Ostrom, E. 2007. *Sustainable social-ecological systems: An impossibility*? Presented at the 2007 Annual Meetings of the American Association for the Advancement of Science, "Science and Technology for Sustainable Well-Being," San Francisco, 15–19 Feb 2007.

Senge, P.M. 1990. *The fifth discipline: The art and practice of the learning organisation*. New York: Doubleday.

Chapter 2
The Changing Meanings of Sustainable Tourism

Stephen F. McCool

Abstract Conventional meanings of sustainable tourism arose out of the convergence of two streams developing in the late twentieth century. One stream arose from the growth of tourism itself and the rising attention society placed on its positive and negative impacts. Another stream originated in the emergent field of international development where a variety of governmental, intergovernmental and non-governmental agencies strove to enhance developing countries' economies. In both streams, this concern evolved into a focus on the environmental and social impacts of development, which lead to the convergence of both streams into the notion of sustainable tourism. And while sustainable tourism involved a conventional meaning, fundamental changes occurring at the very end of the twentieth century and beginning of the twenty-first century have caused us to rethink whether that conventional meaning is still appropriate and useful.

Keywords World Commission on Environment and Development • Environmental impact • Quality of life • Economic impact • Development policy

2.1 Introduction

There is no question that the twenty-first century has brought new challenges and opportunities for tourism development: environmental issues, growing concerns about social justice and income equity, funding of parks and management of visitors, capacity of protected area managers to administer tourism, and expectations of tourism as a panacea for economic and social ills represent the growing and diversifying expectations of tourism. This context of change demonstrates that the meanings of sustainability, sustainable development, and sustainable tourism have also changed and will continue to change as our societies evolve and address these important questions.

S.F. McCool (✉)
University of Montana, Missoula, MT, USA
e-mail: Steve.McCool@cfc.umt.edu

S.F. McCool, K. Bosak (eds.), *Reframing Sustainable Tourism*,
Environmental Challenges and Solutions 2, DOI 10.1007/978-94-017-7209-9_2

This concern is echoed by a chorus of criticism from a number of leading authors. Butler (2013) for example asks if sustainable tourism is the "undefinable and unachievable pursued by the unrealistic" and suggesting it "is surely both a fiction and a smoke screen." Wheeler (2013) christened sustainable tourism a "lame duck" with no discernable purpose. And Sharpley (2009) baptized the concept as a "myth" stating "… a gulf remains between the rhetoric and academic theory of sustainable tourism and the reality of tourism development 'on the ground'". Much of the criticism aimed at the concept reflects an apprehension, expressed primarily by academics (e.g., Weaver 2013) that tourism academics have little impact on the real world of development and management. We believe that this conclusion is true for some, perhaps many academics, but not all. Many of our colleagues are deeply embedded into the practice of development and management.

Liu (2003) and other authors have sometimes responded by indicating that the concept is not the problem, it is that debate and practice have not addressed some fundamental issues very well, such as intra and intergenerality equity, a foundation in the notion of sustainability, or use of systems thinking to more clearly work through goals and processes suggesting that greater clarity can be achieved with more rigorous dialogue. This clarity initially begins with an understanding of where the notion of sustainable tourism comes from, where its roots lie, and why it may be censured for its saliency to twenty-first century needs and thus needs a reframing.

The objective of this chapter, then, is to describe the evolution of the concept of sustainable tourism and argue that conventional meanings are no longer efficacious in addressing the needs of society. While there are many definitions of sustainable tourism that have been put forth, we feel, that given advances in our understanding of social-ecological systems and how they operate, changes in the fundamental assumptions about the character of the world are revolutionizing planning and development, and significant enhancements in the science and practice of tourism, a reframing of the notion of sustainable tourism is not only warranted but is needed to advance tourism development.

The UN World Tourism Organization (UNWTO) has provided a definition of sustainable tourism that encapsulates many of the ideas underlying conventional approaches. The UNWTO expands this definition to include three significant points which in a general sense reflect the conventional framing of sustainable tourism:

1. Make optimal use of environmental resources that constitute a key element in tourism development, maintaining essential ecological processes and helping to conserve natural heritage and biodiversity.
2. Respect the socio-cultural authenticity of host communities, conserve their built and living cultural heritage and traditional values, and contribute to inter-cultural understanding and tolerance.
3. Ensure viable, long-term economic operations, providing socio-economic benefits to all stakeholders that are fairly distributed, including stable employment and income-earning opportunities and social services to host communities, and contributing to poverty alleviation.[1]

[1]http://sdt.unwto.org/content/about-us-5; accessed 15 August 2014.

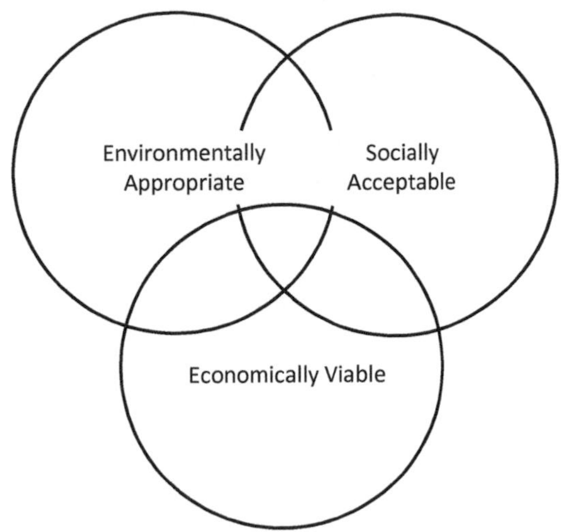

Fig. 2.1 Conventional definitions of sustainable tourism often put it at the intersection of activities (marketing, infrastructure, programs, policies) that are simultaneously environmentally appropriate, socially acceptable, and economically viable

It is difficult to criticize these components on moral and ethical grounds. And who could argue against them. However, sustainable tourism so defined is known as a "guiding fiction". A guiding fiction (Shumway 1991) is a concept that everyone can agree to when expressed at a very general level, and is important in stimulating and organizing social discourse, and thus serves useful social functions. However, guiding fictions tend to fall apart when dialogue gets down to details. For example, what is a "fair" distribution of benefits? What are benefits? How in a rapidly changing world can any action be ensured to be economically feasible in the long term? What is "long-term"? What is "optimal" use of resources? Without appropriate processes to secure a constructive dialogue on such questions, the definition begins to fall apart in terms of its utility on the ground.

One of the most frequent approaches to describing sustainable tourism is tourism development that occurs at the intersection of what is ecologically viable, economically feasible and socially desirable or acceptable (Fig. 2.1), which is at the basis of the UNWTO definition. Conventional definitions grew out of an era where the fundamental assumptions of the world and how it operates—principally in the 1980s and 1990s—are very different than what we assume about and experience the world in the twenty-first century. There is a real question whether these conventional definitions of sustainable tourism are adequate to meet the important challenges in the context, complexity and uncertainty of the twenty-first century.

In this chapter we review the history of the development of the concept of sustainable tourism and propose to reframe this concept to one more suitable to current needs, more salient to the challenges communities confront, and one more useful to the opportunities presented by changing social preferences. At the same time, we respect the goals articulated by UNWTO as a fundamental philosophy to guide all our interactions with others and the environment. We review this history because it not only stimulated an enormous amount of public and academic discourse but also to provide readers a more informed basis upon which to think about reframing the concept.

2.2 Historical Development of the Sustainable Tourism Concept

The history of sustainable tourism is one of two parallel stories, each with several threads, embedded within a larger context of social change, large scale experimentation with development concepts and initiatives, and growing academic interest in tourism. Each story originates in changing social norms and customs, particularly in western society. These stories, that of travel and tourism and that of sustainable development, converged in the very last decade of the twentieth century although the seeds of the convergence sprouted many years earlier. In this section, we outline the development of these stories and discuss the convergence of them with the implications for social policy.

2.2.1 Story 1: Rising Interest in Tourism

2.2.1.1 Tourism as a Social Phenomenon

For all practical purposes, tourism on a massive scale did not begin until after WWII. While the early twentieth century had seen gradual increases in travel, primarily because of steamships, it wasn't until the technology of air travel could be produced in large amounts and at low costs that both domestic and international travel saw dramatic increases. When jet airliners were introduced in the late 1950s with consequent reductions in travel time and a parallel decrease in fares, the demand for travel across the U.S. and to different continents dramatically accelerated, making large scale movements of tourists to distant regions from their homes possible. As well, the decline of conflict between nations—although tempered by the tensions and proxy battles of the Cold War—removed a major obstacle to travel, particularly to places outside Europe and the Americas. Security and safety is the most significant dampening factor on travel and when this limiting factor is removed, travel booms.

While the "Grand Tour" of European destinations, attractions and cities remains exceedingly popular, international travel grew dramatically in the immediate post-war period, quadrupling in the 20 year period from 1950 to 1970 (Fig. 2.2). Domestic travel during this period rose significantly as well, with units of the U.S. national park system becoming so popular with the American public that the U.S. National Park Service embarked on a 10 year program (termed Mission '66) to upgrade visitor facilities to accommodate the increased demand.

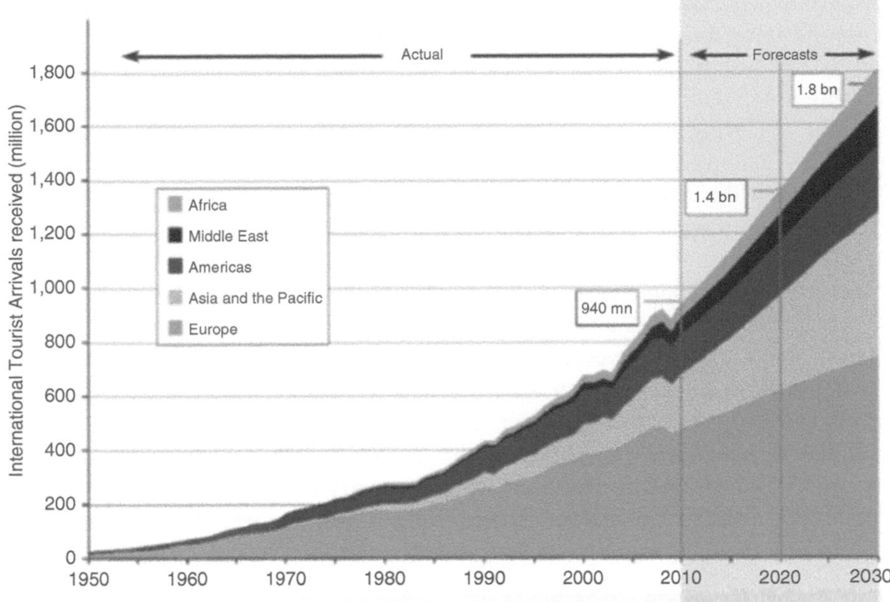

Fig. 2.2 Actual and projected international arrivals (Source: UN World Tourism Organization 2014)

Until the mid 1970s, tourism was perceived primarily as a social phenomenon: it was something everyone did, it was relatively benign in terms of its social and environmental consequences, and it was something that few people even thought about once their vacation was over. Certainly, the study of tourism and recreation as academic subjects were all but unknown in universities. However, during the 1970s, it gradually became recognized that tourism held economic potential and significance, that it could be a tool for employment, income and tax revenue, that if communities, and states advertised (promoted) their "attractions" people would come, spend money and go home, leaving the local residents wealthier. It was during this period that the principal professional tourism research organization in the U.S.—the Travel and Tourism Research Association—was established through an amalgamation of two previous groups and began publishing its periodical the "Journal of Travel Research".

During the 1970s, along with growing travel, the types of destinations and experiences tourists sought began to diversify, with Africa, east Asia and the Pacific beginning to experience rapid increases. At first, the interest in these places was limited to a relatively few hardy souls who liked adventure. While the Pacific region had stabilized with the conclusion of WWII and implementation of recovery efforts, other regions were politically unstable—South Africa for example had an apartheid government which was beginning to unravel—and few contained the infrastructure—such as hotels, lodges, roads, telephone, air travel, safe water—that was needed to support a tourism industry large enough to have much economic impact.

Nevertheless, as travel grew, the few adventurous souls became the many. With the implementation of democracy in South Africa, that country, and the nearby nations of Botswana and Namibia also became important destinations offering, along with such countries as Kenya and Tanzania, wildlife-watching safari adventures. During the 1980s, wildlife watching had become a very popular activity in the U.S., and travelers sought foreign destinations, particularly in the tropics and Africa that held a diversity of animal and birdlife. Travel no longer was dominated by the Grand European Tour, but increasingly became a niche industry, with a large variety of destinations, a diversity of settings, and a variety of activities.

In the 1980s, though, as travel to developing countries increased, particularly in poverty stricken areas and into relatively pristine environments, planners, academics, providers and travelers came to increasingly question large scale tourism development, and began to plan, seek and operate smaller scale resorts and lodges that seemed more attuned to their social and biophysical surroundings. This movement gradually became known as "ecotourism"—tourism that is small in scale, "benefits" the resource, involves the local community, and has an educational focus. Lodges in places such as central America and Africa that were small, but offered high quality facilities, cuisine and interpretation were increasingly sought.

Tourism as a social phenomenon lead to many benefits, both for the tourists and residents of host communities. It has resulted in a more enlightened and educated world, it has been a force for international understanding and nurturance of cultural relationships, and it improved appreciation of the circumstances, for tourists visiting developing countries, of the depth and distribution of poverty, and driven greater awareness of environmental values. And yet, these very positive values have not been without cost: tourism, as we shall see also carries with it the potential for some significant consequences to the very cultures that tourists come to see and understand.

2.2.1.2 The Economic Significance of Tourism

Tourism has been more than simply masses of people visiting different and foreign destinations, it has had economic consequences as well. The growing recognition in the 1970s of the significance of tourism as an economic development tool blossomed in the 1980s. Two particular developments were important in this transformation in the U.S. First, the downhill ski industry developed massive scale resorts, particularly in Colorado, then Utah, British Columbia, and of course Europe. These resorts brought with them the need for hundreds, if not thousands of employees. Economic development groups quickly saw the potential of larger scale tourism in countering some of the economic ills that struck the U.S. in the 1980s, following the depressing economic effects of the oil embargos in the 1970s.

Second, region wide economic restructuring was occurring in the western U.S. Communities whose economic foundation had centered on natural resources—agriculture, mining, timber, forage—were experiencing mill closures, mining turn downs, and uncertain commodity prices. All of which increased unemployment and the resulting negative social consequences. There were several causes to this change. One was simply increased overseas competition. Another was international exchange valuation. Another, and major factor, was changes in American attitudes toward the environment, which lead to an increased contentiousness over the logging and mining policies on the federally administered public lands of the west. These policies in the 1980s had initially favored resource extraction, but public opposition quickly jelled, with the result that supplies of commodities that were available not only declined, but their predictability became increasingly uncertain. Manufacturing closures increased.

The 1980s were a period for some states, such as Montana, that were close to a depression, with high unemployment rates, and net outmigration. The result was a search duplicated many times, conducted by communities to see if there was any other economic base industry that could substitute. For many, they turned to tourism. And tourism has become big business in most states in the U.S. and in many foreign countries. The World Travel and Tourism Council (2012) estimates that travel and tourism total direct, indirect and induced economic impacts total over US $6.3 trillion, and account for 9.1 % of the world's Gross Domestic Product.

While these statistics are impressive, tourism as an economic development tool remains susceptible to arguments that many of the jobs (such as housekeeping, waiting and other services in hotels and lodges) are low wage and dead-end, lacking potential for building skills and upward mobility. This is often countered by the arguments that these jobs provide opportunities for members of society at the lowest level of skill to be gainfully employed, that the induced effects of tourism spending lead to higher paying service jobs supporting tourism (attorneys, accountants) and that there is a lot of opportunity for entrepreneurship.

The net consequence of travel and tourism is an accelerating importance of the industry to local, regional and national economies, leading to increased economic dependency. In terms of sustainability, increased dependency means reduced resiliency to disruptive forces (natural disasters, economic downturns, terrorism). We will turn to the notion of resiliency later.

2.2.1.3 Cultural and Environmental Consequences of Tourism

The flow of tourism, particularly to developing countries, is not without its significant social, cultural and environmental consequences. These countries, not being developed, often can be characterized as containing traditional cultures, vulnerable to secular values, behaviors, and perspectives brought by tourists and also susceptible to disease and disaster. Often, these cultures were dominated by a subsistence lifestyle and the onset of tourism facilitated a conversion to a cash economy, for which they were ill-prepared, culturally, socially and politically. As tourism in these places increased, as a result of genuine curiosity and a desire to learn, local communities began to feel they were being exploited and that their culture was being lost and transformed into a western one that was inconsistent with previous values and orientations.

These concerns can be described as: (1) tourism disrupted, and eventually commodified local cultures and their icons, events and values; (2) tourism resulted in unacceptable impact to the environment that was often the basis for the industry; and (3) local communities did not have access to the decisions that lead to tourism development (including promotion, facilities and infrastructure); and (4) new facilities often did not employ members of the local community. These concerns grew into others, such as perceptions of power differentials between tourists and hosts, dominance of foreign ownership and management of private lodges and facilities, development of human trafficking and sex tourism, and feelings of inequity within communities as some people benefited from tourism but others did not.

These arguments about tourism were significant, and in the 1980s, the response was development of a new form of tourism referred to as "ecotourism". Hector Ceballos-Luscarin was one of the first to describe ecotourism noting in 1983 that it is:

> That form of environmentally responsible tourism that involves travel and visitation to relatively undisturbed natural areas with the object of enjoying, admiring, and studying the nature (the scenery, wild plants and animals), as well as any cultural aspect (both past and present) found in these areas, through a process which promotes conservation, has a low impact on the environment and on culture and favors the active and socioeconomically beneficial involvement of local communities.[2]

The notion behind ecotourism would be that tourists visit pristine areas and local communities in indigenous cultures in small groups, interact with the local community to learn, appreciate and become more sensitive to cultural values, spending would benefit the community and any parks or protected areas tourists visited, and they would leave little environmental impact. This notion has gradually diversified somewhat to include support for human rights, labor agreements and with greater focus on income redistribution in the form of "pro-poor tourism".

The concerns about these consequences remain. Indigenous peoples in particular remain skeptical of tourism often considering it just another plot by western, industrialized countries to continue to marginalize their cultures and rights. This has been exemplified in conservation by the notion of "fortress conservation" in which the designating national parks and protected areas in a number of developing countries resulted in the removal of indigenous peoples from these parks, and the loss of their subsistence livelihood. This process has become the center of a major global level debate for the last quarter century.

2.2.1.4 Concern About the Environmental Consequences of Tourism

As these concerns were developing, there were parallel growing apprehensions about the status of the globe's biodiversity. A number of authoritative non-

[2]The 1983 date is based on an interview with Ceballos-Luscarain published in 2000 and cited in http://www.planeta.com/ecotravel/weaving/hectorceballos.html

governmental organizations such as IUCN, Conservation International, World Wildlife Fund, The Nature Conservancy and Birdlife International pressured for more designated protected areas, but at the same time both promoted tourism and worried about its impacts on biodiversity. At such conferences as the World Parks Congresses in 1992 and 2003, the Global Congresses on Tourism and the Environment (1992, 1993 and 1994) and at other meetings, delegates often debated the impact of tourism on biodiversity. So, on the one hand, tourism was viewed as a tool for economic development, as a source of funding for management and as an instrument for environmental protection while on the other, there was growing concern about the cultural and environmental footprint of tourism.

Many of these concerns were articulated in the form of proposals for "alternative tourism"—using ecotourism as a model, a number of academics and activists proposed new kinds of tourism development that emphasized small scale, low impact type of facilities and activities. In the academic sectors, the social concerns were most dramatically depicted by Doxey's (1975) irridex and Butler's (1980) resort life cycle. Doxey proposed that visitor-resident attitudes evolve through four stages as the amount of tourism increases—getting to the point where residents hold feelings of annoyance and eventually antagonism toward tourists. Butler hypothesized that destinations go through predictable stages—growth, maturity, decline, for example—which are partly a function of resident responses to growing tourism.

2.2.2 *Story 2: Sustainability—Arose from Initially from Four Separate, but Eventually Converging Themes*

The rise in interest in the concept of sustainability has its source in four evolving and parallel ideas, generally having roots in nineteenth century authors, but not coming to fruition until the second half of the twentieth century.

2.2.2.1 Impact of Human Activity on the Environment

During the 1950s, there was little if any recognition of the impacts of human activity on the environment. While there were various initiatives and concerns expressed, such as the debate over a Wilderness bill proposed in the U.S. Congress, for example, that would set aside lands protected from development, those efforts were largely isolated events. That changed, however, in the U.S. in 1962 with the publication of the book *Silent Spring* by Rachel Carson, who was a marine naturalist and an author of immensely popular books on the natural history of the seas and oceans. In *Silent Spring*, Carson (1962) documented the impact of the chemical DDT—used extensively at that time as an agricultural pesticide—on the food chain. The book's assertions came to prominence when President Kennedy ordered his Science Advisory Committee to investigate her claims. The public awareness resulting from *Silent Spring* was a key step in generating societal concerns about the environment.

The infant environmental movement of the 1960s, was given more impetus by changes in national policy occurring at the same time—namely increased interest in protecting national parks, negative reactions to construction of dams (articulated in the Wild and Scenic Rivers Act of 1968), protection of wildlands through the Wilderness Act of 1964, and, ultimately in the National Environmental Policy Act of 1969. Many other countries followed with similar legislation and rules. Environmental analyses are now part of nearly every nation's legislative agenda and policy to protect the environment.

These formal policy statements were complemented by several reports and articles. Garrit Hardin's (1968) essay, *The Tragedy of the Commons*, published in 1968, triggered a close examination of the impact of institutional structures on environmental protection and management.

Growing concern about environmental impacts of development was addressed in the 1972 United Nations Conference on the Human Environment (United Nations 1972). The Conference proclaimed in part:

> The protection and improvement of the human environment is a major issue which affects the well-being of peoples and economic development throughout the world; it is the urgent desire of the peoples of the whole world and the duty of all Governments.

An outcome of the Conference was the creation of the United Nations Environmental Programme, headquartered in Nairobi, Kenya. The Club of Rome report (Meadows 1972), also released in 1972 and titled *The Limits to Growth* pointed out that not only could the rates of consumption of non-renewable resources not be sustained into the future, but that piecemeal approaches to solving resource degradation, supply and management would not be successful. While the Club of Rome report continues to be hotly debated, it never-the-less triggered global level debate about the sustainability of human consumption patterns.

Many of these concerns were crystallized with the publication of the *World Conservation Strategy* in 1980 by the International Union for the Conservation of Nature (IUCN). The strategy recommended there, and the follow-on 1991 document *Caring for the Earth*, recognized the need to integrate and conservation and poverty alleviation. While these publications were widely distributed globally, they did not have much impact in North America.

2.2.2.2 International Interest in Development

Following WW II, the U.S. and many other nations embarked on significant foreign aid programs, most aimed at developing countries. The fundamental aim of these aid programs (although somewhat colored by strategic objectives during the Cold War era) was to implement large scale projects to increase income, move to cash economies, remove poverty, treat resource degradation, and broaden markets for developed world products.

The investments in large scale capital development (e.g., dams, transportation, power) were substantial. Gradually, it became clear that poverty and resource degradation were linked (although the causes may have been deeper than simplistic notions that poverty leads to degradation), that large scale projects often did little ultimately to address the root causes of poverty, that such projects often lead to even more inequitable income distributions, and that the projects themselves often held irreversible environmental consequences.

During the late 1970s onward, academia and activitists increasingly critiqued various projects suggesting that many never adequately addressed the systemic causes of poverty and often questioned the capitalistic model they employed. While the efficacy of the specific arguments varied, the result was a broadly recognized conclusion that resource development projects lead to consequences that were usually opposite of what was intended and frequently unanticipated, that large scale projects that did not include local communities in their design and implementation lacked the social and cultural legitimacy needed for successful implementation, and that conservation and poverty alleviation needed to be integrated in ways that hadn't yet been attempted.

2.2.2.3 Growing Interest in the Notion of "Quality of Life"

One of the outcomes of international development aid was a rising interest in the notion of quality of life. A number of individuals began a critique of many monetary measures of national well being (e.g., per capita income, gross domestic product). The basis of this critique was that such measures do not include things that are important to people, such as access to education and health care, clean environments, and so on. Their arguments suggested that traditional measures of wealth were too narrow, even conceptually invalid, that one must consider the quality of life in development project proposals. Thus, that while a project may have been oriented toward poverty alleviation if it did not include proposals, developments and so on say to raise access to water for domestic purposes, it did not really make much progress.

This critique called into question the fundamental purposes of economic development—that it wasn't income or gross domestic project that should be the objective, but rather the things that people need for a better quality of life. As this critique grew, it came to include concerns about the protection of important cultural values, preservation of languages, women's rights and roles in society, and environmental and social justice.

2.2.2.4 Changes in Models of Governance

The growth in interest in quality of life and the often inequitable consequences of large scale development activities converged to produce interest in natural resource governance systems. In the U.S., this was stimulated in the 1960s by a provision in the Wilderness Act requiring public participation in recommendations for wilderness designation and by the requirement in the National Environmental Policy Act for public engagement during the scoping process for a decision and following publication of draft alternatives, and finally by accelerating discontent with top-down, command and control planning processes. Contributions by individuals such as John Freidmann (*Retracking America*), Sherry Arnstein (*Ladder of Citizen Participation*), Julia Wondolleck (*Public Lands Conflict and Resolution*), Dan Kemmis (*Community and the Politics of Place*), and Daniel Yankelovich (*Coming to Public Judgment*) provided a further critique of extant public participation processes from the 1970s to the 1990s.

This growing interest in more horizontal, upfront citizen participation was discussed under varying rubrics, such as power-sharing and co-management. This appeal was often linked to sustainable development because of the equity issues that sustainability implies—one must be involved in a decision to properly identify the consequences of it. This interest, in the words of Dan Kemmis represented a "living critique" of traditional the scientifically based expert driven models of planning and management that grew out of the Progressive Era philosophy of public decision making. This era placed a premium on expertise in making decisions in order to remove politics from them, but a consequence was to marginalize public input and non-technical forms of knowledge. It continues to dominate natural resource decisions today, although there are numerous attempts to reform this paradigm (e.g., Kohl and McCool 2016).

With growing unrest about the marginalization of the poor, minorities and women in formalized natural resource decision making, it became clear that a sustainable world was one that was also much more equitable and inclusive in policy development and implementation. How benefits from various resource management activities, such as tourism, are distributed became important social justice oriented questions.

2.2.2.5 The Convergence

The four threads began converging in the 1980s, when it was increasingly recognized that economic development was too narrow a definition of what the task was, that participation in decisions affecting people was fundamental to social and political acceptability of policy and projects and that maintaining options for future generations required an environment whose quality and biodiversity was at least as great as the one current generations held. The threads were encapsulated and integrated and then brought to international attention in 1987 with publication of the Brundtland[3] Commission report *Our Common Future*. This commission, more formally named the "World Commission on Environment and Development" had been chartered in 1983 by the UN to study how development could address environmental issues, trigger international cooperation on these issues, motivate increased cooperation to achieve sustainable development, and develop a long term global agenda for environment and development. Gro Brundtland, the Prime Minister of Norway headed the Commission, noted in the report that initiating the study represented "a clear demonstration of the widespread feeling of frustration and inadequacy in the international community about our own ability to address the vital global issues and deal effectively with them."

While the notion of sustainable development had begun to surface prior to the Brundtland Commission report, it was this book that crystallized global interest in a new approach to development. The Commission defined sustainable development as "development that meets the needs of the present without compromising the ability of future generations to meet their own needs". While this statement seems rather straight forward, it has been debated ever since the report was published. How can development be sustainable? Are the interests and preferences of the current generation and future generations similar? What are needs? What are desires? What is it that we should be sustaining? These questions and many others resulted from this report.

The report was perfectly timed to stimulate debate and dialogue about sustainability and institutional response at the UN Conference on Environment and Development held at Rio de Janeiro, Brazil in 1992. Commonly termed the "Earth Summit", a principal output was Agenda 21, a set of proposed national, regional and even local policies to achieve sustainable development goals. Agenda 21 has been hotly debated in some places and in others adopted with little fanfare.

2.3 Integrating Sustainable Development and Tourism

During the late 1980s' and especially following the Rio Conference, academics, planners and activists began considering applying the notion of sustainable development to tourism. This idea crystallized in the early to mid-1990s. Reflecting this, a new journal, The Journal of Sustainable Tourism was launched in 1993. The earliest use of the term "sustainable tourism" in the title of a scientific article or publication occurred in 1990 by J.J. Pigram with the First World Conference on Sustainable Tourism held in 1995 in Spain. While these dates seem relatively recent, the seeds for the term "sustainable tourism" were planted early, with the concerns about the size and scope of tourism development, with increasing awareness of social and environmental impacts, and with debate about the types of jobs tourism

[3]Gro Brundtland was Prime Minister of Norway at the time. The Commission is commonly and informally referred to as the "Brundtland Commission".

created as we have mentioned. The concern about sustainable tourism, and its integration within the broader concept of sustainable development could be viewed as a maturation of the concept: that through debate and discussion, at all levels and sectors, the linkage of sustainable tourism into broader concerns about the environment, the economy and social systems represents a more sophisticated view of the problems facing humanity.

There is no question that academics, development specialists and activists found the notion of sustainable tourism appealing, if for nothing else it provided an alternative to a market-driven, capitalistic based and narrowly defined notion of tourism and its goals. Sustainable tourism appeared to provide greater opportunities for minimizing the growing environmental impact of tourism development, for enhancing employment and building skills among low income peoples, and as a means of achieving larger scale goals concerning generation of income that could be directed toward management of sensitive areas of natural heritage.

The popular paradigm of sustainable tourism occurring at the intersection of what is ecologically appropriate, socially and culturally acceptable and economically feasible became the mantra underlying literally thousands of developmental experiments across an increasingly contentious globe. While peace seemed to be at hand in the mid 1990s, the 9/11 attacks in 2001 changed the world forever, giving rise to wars, increased terrorism, accelerating concern about security and greater instability in a variety of locations that would otherwise be attractive locations for tourism development. These changed circumstances pointed out that the world was actually much more contentious and uncertain than assumed in earlier discussions about tourism and its utility as a development tool. This has lead to some questioning the assumptions underlying sustainable tourism and the simplistic approach to it developed in the first decade of its application. For example, what is economically feasible is highly dependent on market connectedness, which is influenced by access and regional security and interest rates which vary significantly over time. What is culturally acceptable changes over time and is influenced by the dominant political-religious system in a number of countries. And what is ecologically appropriate depends on the character of the local ecosystem as well as how we see that system—one that is stable or what that is changing. Too, the notion of appropriateness or acceptability of impacts is one culturally and socially driven, but influenced by science.

That said, the dialogue about sustainable tourism and its various approaches and implications has triggered many experiments, studies, policies and strategies focused on reducing the negative social and biophysical impacts of tourism as well as enhancing its positive consequences. The literature contains thousands of case studies examining sustainable tourism at the community level, a few of which are included in this book; more thousands of discussions are held each year about what sustainable tourism means and how it can be implemented. Notwithstanding the critiques of Weaver and Sharpley briefly noted earlier in this chapter, many academics are involved in these experiments, studies, assessments and dialogue. Policy makers have been exposed to the notion of sustainable tourism, and while we might critique them in many cases for not acting accordingly, at least some have.

Tourists have become more sensitive to the environment, the unique qualities of cultures and the contribution of their spending to a better quality of live for those most vulnerable. This discourse has had more positive impact than if it had never occurred and it is through experimentation, monitoring, assessment and reflection that we learn.

2.4 Some Confusion About the Meanings of Sustainable Tourism

And our learning has exposed a number of challenges and opportunities in the notion of sustainable tourism. As we argue here, conceiving sustainable tourism as that which happens at the convergence of ecological viability, social acceptability and economic feasibility simply no longer provides the utility it once had. We live in a complex world, one where decisions made 12 time zones away from us today can determine whether we work tomorrow, where incremental changes made in policy can lead to dramatic consequences, and where uncertainty makes predictability something we yearn for.

We have noticed that the idea of sustainable tourism is used in at least three distinct ways, and these meanings often make discourse confusing. The first meaning involves sustainable tourism as a small, sensitively developed and managed enterprise catering to the needs, especially food, lodging, and guiding services, of tourists interested in nature and culturally based experiences. This notion we call the "small is better" idea, which has great appeal among many because they believe it counterbalances the much larger scale tourism development that often characterizes beach resort destinations such as Cancun, Honolulu, Bali and so on. Smaller scale businesses do have less impact, obviously, than larger scale businesses, but there is another question about efficiency of scale, that perhaps the per capita impact of smaller scale businesses may be greater than the larger scale ones.

Another meaning that we often hear at meetings is that sustainable tourism means sustaining small businesses or the local tourism industry. That is, sustainable tourism takes on a set of policies to ensure that small businesses survive over a longer term, that they continue to serve tourists, and it becomes a pillar for achieving a stable economic base for the community. In this sense then, when we speak of sustainable tourism we are not necessarily focused on scale, but rather business and financial outcomes of tourism development.

A third meaning is one which we don't hear very often, but we believe is at the heart of the dialogue about sustainable tourism, and that meaning emanates from the question "what is it that tourism should sustain"? Thinking of sustainable tourism in this way requires us to dive deeper into the fundamental purpose of development. Some people may respond to this question in terms of jobs and personal income. Others may say tax revenue. Still others, seeing a more connected world, may argue that tourism provides access to health care and educational opportunity. And still others may see sustainable tourism as a tool to buffer communities against the forces (e.g., the financial sector, disease, war and conflict) that greatly influence the viability of the community itself.

We raise this discussion not to argue that one meaning is more useful than another but to argue for clarity in discussion that otherwise will be characterized by ambiguity and the illusion of agreement. Different meanings carry with them implications for other aspects of policy, planning and management. For example, what indicators would be appropriate to monitor depends largely on what approach to sustainability and sustainable tourism is chosen. Incentive policies, cooperative agreements and governance would be impacted differently by different meanings. And we also suggest the differences in meanings reflect a growing, uncertainty about the usefulness of the concept of sustainable tourism, at least in the conventional sense.

And clearly, given the focus on sustainable tourism in many geographies as a tool for development, the question of what tourism should sustain would impact policy and development goals. By considering this question, we raise the possibility of assessing tourism not as a specific kind or size of business, type of tourist activity or motivation of visitors, but rather as a means of achieving other socially worthy goals as well as integrating tourism into deliberations about a community's economy and development as a whole.

2.5 Emergence of Alternative Approaches to Sustainability and Sustainable Development

The Brundtland Commission report not only stimulated worldwide dialogue on the direction and purpose of development, raised important issues and questions about poverty and its alleviation, and moved policy to be more engaged in considering inequities in access to basic human needs, such as health, education, and food security, but the report also stimulated other ways of thinking about sustainability and sustainable development.

Jacob (1994) indicates that sustainable development (in the Brundtland Report context) "retains the Western tradition's assumptions that humans have a privileged role in the biosphere" while the followers of Deep Ecology would question whether sustainability can be achieved in a Western society where nature is viewed as having value for human purposes and no intrinsic worth itself. Commodification of nature is a fundamental roadblock in the pursuit of sustainability. Robinson (1993) has criticized sustainable development (in the Caring for the Earth context) as being "simplistic" and containing an almost exclusive focus on human beings. Adams (1990) suggests that "ecocentric" approaches to sustainability "demand fundamental change in political economic structures" and are not simply about reforming resource management policy.

These comments echo Martin O'Conner's (1994) stinging rebuke of the ability of capitalism to achieve sustainability goals. He argues that "it is always the capitalist system that is to be reproduced and sustained, not any individual capital" such as nature, specific ecosystems, forests or parks. Thus, capitalism will use up/tradeoff natural and human capital in its pursuit of maintaining capitalism as a social norm because of its allegedly "predatory" and "exploitive" practices. While capitalism can be severely critiqued for its preference for the short-term, lack of attention to externalities and effects on the disenfranchised, alternatives are few. Socialism, as revealed following the fall of the Soviet empire, did no better in protecting the environment, providing a high standard of living and considering the long-term consequences of economic development.

If sustainability involves intergenerational and intragenerational equity or bequests of capital, then we would argue that fundamental relationships among people and between generations have been changed. Such redistributions and bequests involve changes in power relationships. Future generations have been given an increased voice. The disadvantaged now speak with authority while the advantaged must listen. While power is not necessarily something people give up voluntarily, by implementing sustainability society has apparently made the decision that existing power relationships are not beneficial in the long-term, that a moral people give up wealth so that others may enjoy some increase in benefits, and that there is powerful ethical obligation of a civilized society to those coming later. As Adams (1990) observes, such changes are not simply reforms, but reflect a radical view of sustainability: "'Green' development is not about the way the environment is managed, but about who has the power to decide how it is managed".

While these viewpoints are not necessarily widely held (and it could be argued that a democratic government is a result of their application), the institutional capacity to shift power to the future especially may not as yet exist. The shift in power is focused not on ensuring income for future generations, but instead on allowing those generations voices in how the environment will be managed in the present. This obviously presents severe obstacles, because no one has been designated to represent those preferences, and current generations have little idea of the preferences of future generations.

These critiques suggest that there is more to the notion of sustainable development and sustainable tourism than conventional thinking suggests. They imply a number of issues addressing power, scale, purpose and ethic that are deeply engaged in tourism development particularly when the term "sustainable tourism" is raised.

2.6 The Issue of Scale

Surely, one of the most fundamental tasks of sustainable tourism policy is to develop some means of assessing progress toward achieving our goals. Measurement provides us with the necessary feedback for learning and reflection and to correct actions and ensure that we are on the pathway to sustainability. The scale at which sustainability is measured is an important, though often neglected component of sustainability discussions. Lee (1993) notes that mismatches between human and biological scales leads to unsustainable resource uses. There are at least four types of scale that are relevant to development of sustainable tourism policies. First, there is the question of appropriate temporal scale: over what period of time do we judge the sustainability of an industry, community or ecosystem? There may be conflict over tourism activities that aim for intergenerational equity and management directed toward intragenerational equity (Dovers and Handmer 1993). Dixon and Fallon (1989) state that "the shorter the time horizon [in resource decisions], the less likely any pattern of resource use will be sustainable over long periods of time" and they ask "How far into the future should we worry about?" Mismatches between temporal scales leads to one generation bearing the costs of another generation's benefits.

A second type of scale is spatial: over what spatial scale is sustainability measured? Mismatches between spatial scales can lead to some people, communities or ecosystems bearing the costs without associated benefits. Can sustainable tourism in one locality come at the expense of sustainability for another community, as Colin Hunter (1995) argues? Multiple spatial scales are involved in sustainability decisions: Could sustainable tourism development at the local scale lead to unsustainable tourism patterns at the regional scale? Focusing sustainable tourism efforts solely at the local level may lead into another type of wicked problem: policies and their implementation may be doing well, but in a globalized economy decisions made distant from the community may affect the achievement of a sustainability goals.

People interact over varying social organizational scales, such as families, neighborhoods, communities and so on. Addressing the social organizational scale helps us understand for whom sustainable tourism is being developed. Lee (1993) argues that a fourth scale mismatch occurs, what he terms a functional scale mismatch. Functional mismatches occur because natural systems, such as marine environments, are complex, but human actions and institutions are necessarily specialized. Achieving a specialized goal, such as tourism development may conflict functionally with the sustainability of a particular ecosystem.

Institutions have enormous influence over the ability to develop sustainable tourism. Institutions developed in the western United States to deal with resource management were derived primarily in an era based on a resource utilization philosophy. The instrumental philosophy governing resource management in this era resulted in highly specialized institutions designed to enhance efficiency of use of individual resource commodities. Many institutions lack the flexibility to address new problems and new challenges. They are not particularly well designed for sustaining over long periods of time entire ecosystems or communities. A second issue with respect to tourism concerns the institutional orientation of many tourism development agencies, particularly in the United States. Many state level agencies are solely concerned with promotion of tourism, do little to assess negative social and environmental impacts, and rarely monitor the outcomes of their promotional programs, such as effects on employment, quality of life and protection of the natural and cultural heritage.

2.7 Conclusion

Aside from the conventional approaches to sustainable tourism there are an emerging set of arguments, such as the redistribution of power to future generations, the use of sustainable tourism to build local community resiliency, provide a constellation of viewpoints influencing decisions about tourism development These set of emerging perspectives reflect to some extent the dissatisfaction with the conventional approach that we have articulated above. We propose that we need more discussion about the purposes of sustainable tourism, who it benefits and who it doesn't, how to operationalize it to ensure it does not become a guiding fiction and how to enhance our thinking about sustainability to be more reflective and considerate of implications of policy and proposals.

The focus in this book is sustainable tourism occurring at the community level. We have chosen this emphasis because it not only is at this level where tourism occurs, but this scale is where the action is, where development occurs and where people interact to get things done. National governments can facilitate tourism development through enabling policies and legislation, through subsidies, through large scale infrastructure—such as airports and highways—and through other means, but it all comes back to the community level.

Being keenly aware of the scale issues mentioned above, at the community level we translate the abstract into the real, we move beyond policy to action, whether it be land use zoning, capacity building, collaborative marketing, or cooperative production of arts and crafts. In this book, we focus on tourism as a means of development, which we more closely articulate in the next chapter and the following sections. This tourism development would have as a goal to build community resiliency in the face of the uncertainties and complexities of the twenty-first century.

References

Adams, W.M. 1990. *Green development: Environment and sustainability in the third world*. London: Routledge.

Butler, R. 1980. The concept of a tourist area cycle of evolution: Implications for management of resources. *Canadian Geographer* 24(1): 5–12.

Butler, R. 2013. Sustainable tourism—The undefinable and unachievable pursued by the unrealistic? *Tourism Recreation Research* 38(2): 221–226.

Carson, R. 1962. *Silent spring*. New York: Houghton Mifflin Harcourt.

Dixon, J.A., and L.A. Fallon. 1989. The concept of sustainability: Origins, extensions, and usefulness for policy. *Society and Natural Resources* 2(1): 73–84.

Dovers, S.R., and J.W. Handmer. 1993. Contradictions in sustainability. *Environmental Conservation* 20(3): 217–222.

Doxey, G. 1975. A causation theory of visitor-resident irritants: Methodology and research inferences. In *Travel and tourism research association sixth annual conference proceedings*, San Diego.

Hardin, G. 1968. The tragedy of the commons. *Science* 162(3859): 1243–1248.

Hunter, C.J. 1995. On the need to re-conceptualise sustainable tourism development. *Journal of Sustainable Tourism* 3(3): 155–165.

IUCN. 1980. *World conservation strategy*. Gland: World Conservation Union, United Nations Environment Programme, Word Wide Fund for Nature.

IUCN. 1991. *Caring for the earth: A strategy for sustainable living*. Gland: IUCN.

Jacob, M. 1994. Sustainable development and deep ecology: An analysis of competing traditions. *Environmental Management* 18(4): 477–488.

Kohl, J., and S.F. McCool. 2016. *The future has other plans*. Denver: Fulcrum.

Lee, K.N. 1993. Greed, scale mismatch, and learning. *Ecological Applications* 3: 560–564.

Liu, Z. 2003. Sustainable tourism development: A critique. *Journal of Sustainable Tourism* 11(6): 459–475.

Meadows, D.H. 1972. *Club of Rome. The limits to growth; a report for the Club of Rome's project on the predicament of mankind*. New York: Universe.

O'Connor, M. 1994. On the misadventures of capitalist nature. In *Is capitalism sustainable? Political economy and the politics of ecology*, ed. M. O'Connor, 125–151. New York: Guilford.

Pigram, J.J. 1990. Sustainable tourism-policy considerations. *Journal of Tourism Studies* 1(2): 2–9.

Robinson, J.G. 1993. The limits to caring: Sustainable living and the loss of biodiversity. *Conservation Biology* 7(1): 20–28.

Sharpley, R. 2009. *Tourism development and the environment: Beyond sustainability*. Oxon: Earthscan.

Shumway, N. 1991. *The invention of Argentina*. Berkeley: University of California Press.

United Nations. 1972. *Declaration of the United Nations conference on the environment*. Nairobi: UN Environmental Program. http://www.unep.org/Documents.Multilingual/Default.asp?documentid=97&articleid=1503

UN World Tourism Organization. 2014. *Tourism highlights: 2014 edition*. Madrid: UNWTO.

Weaver, D. 2013. Whither sustainable tourism? But first a good hard look in the mirror. *Tourism Recreation Research* 38(2): 231–234.

Wheeler, B. 2013. Sustainable tourism: Milestone or millstone? *Tourism Recreation Research* 38(2): 234–239.

World Commission on Environment and Development. 1987. *Our common future*. Oxford: Oxford University Press.

World Travel and Tourism Council. 2012. *The comparative economic impact of travel and tourism*. Oxford: Oxford Economics.

Chapter 3
Tourism, Development, and Sustainability

Keith Bosak

Abstract This chapter approaches sustainable tourism by first conceptualizing tourism as a form of development that depends on conservation of the resources that attract tourists. With this lens, sustainability becomes a top priority as tourism destinations attempt to maintain and enhance their attractiveness to tourists. Concurrently, tourism is an economic activity and is a driver of and is driven by global capitalism. As such, there exists a tension between tourism and sustainability. That tension is explored in this chapter through the question: What is to be sustained? The answer to this question is explored in the context of socio-ecological systems and ideas of resilience. The chapter concludes with a discussion of how re-framing sustainable tourism could be implemented through a set of tools.

Keywords Tourism • Sustainable development • Capitalism • Socio-ecological systems • Resilience

3.1 Introduction

In this chapter, tourism is treated as a form of development that also depends on the conservation of the resources that provide the attraction for tourists. With this perspective, sustainability becomes paramount in order to maintain tourism and the environments (natural and built) on which the activity depends. At the same time, tourism is also an economic activity, one that is embedded within and a driver of global capitalism (Fletcher 2011). Tourism provides economic opportunities and a chance for people in even the remotest areas of the world to engage with the global economy. At the same time, tourism is more complex as it is often simultaneously used to forward goals of poverty alleviation and empowerment of marginalized people. Therefore, tourism also has social, political, and justice components (Weaver 2001; Hipwell 2007; Bosak 2010; Schellhorn 2010). In this chapter, we explore the various components of tourism in the context

K. Bosak (✉)
University of Montana, Missoula, MT, USA
e-mail: Keith.Bosak@umontana.edu

S.F. McCool, K. Bosak (eds.), *Reframing Sustainable Tourism*,
Environmental Challenges and Solutions 2, DOI 10.1007/978-94-017-7209-9_3

of development and sustainability in order to illustrate the complexities involved in developing and maintaining sustainable forms of tourism. The chapter concludes with a discussion of how sustainable tourism might be re-framed in order to be more effective and some tools are introduced that could assist in the process.

3.2 Tourism as a Path to Development

Because tourism is often seen as a path to development and sustainable tourism is envisioned as a path to sustainable development, it is important to understand what we mean by development and how tourism is often used (particularly in developing countries) as a path to development. Development can be defined in various ways and in this chapter we will explore development through several different definitions. Most simply conceptualized, development can be seen as a process of growth and or transformation. This can occur in an economic or non-economic sense. Economic development can be defined most simply as progress in the economy. The goal of this progress is higher standards of living. We can also think of development in a social sense. Social development entails the transformation of social structures with the goal of benefitting people. In thinking about what development looks like it is helpful to define development as a change in land use from one activity to another more economically productive activity. We have all witnessed this: farmers who can no longer afford to keep their farmland sell to developers who change the use of that land to something more economically productive, perhaps a housing development or strip mall.

When we think of tourism in the context of development, it becomes clear that tourism is itself a form of development. Tourism as an economic activity is based in the capitalist mode of production and in neoliberal economics. The goal of tourism as economic development is to grow local, regional or national economies through income gained by providing tourism products and experiences. Oftentimes neoliberal policies are used to promote this development of tourism. Foreign direct investment is a good example of this and has been used extensively in places like Mexico, Costa Rica and Chile to grow their tourism economies. While tourism as economic development often has benefits at various scales, it is often those who already have capital to invest in tourism enterprises that benefit the most. This leads to what we might call uneven development of tourism and has spurred a trend in thinking of tourism as a form of social development (Britton 1982).

Tourism as social development is concerned with the well-being of people, particularly the poor. In this case, ideas of equity and social justice become prevalent and the main goals are poverty alleviation and empowerment of marginalized groups (such as women and minorities) (Bosak 2010). However, we must also remember that tourism is an economic activity and that income generation is a key component of tourism as both economic and social development.

No matter what the focus of tourism (economic or social development or both) it is also important to re-visit what tourism development 'looks like'. This takes us back to the idea of development as a change in land use from one activity to another more productive activity. As tourism develops in an area, tourists begin to require more and more infrastructure, necessitating the conversion of land from activities such as agriculture to restaurants, lodges and gift shops. What this illustrates is what I call the defining characteristic of tourism: That it fundamentally changes the places where it occurs to meet its own needs. What this means is that either form of tourism as development (economic or social) produces impacts both positive and negative. It is also important to emphasize that tourism produces change, a fact that is often overlooked in sustainable tourism (more on that later). Impacts produce change and many impacts from tourism are unanticipated, particularly negative impacts. Because impacts from tourism produce change in environments and social systems, the character of destinations is almost always altered by tourism development and this produces a difficult situation: Tourism is a form of development that depends on the conservation of resources that attract people. Because negative tourism impacts threaten the very existence of the activity in a destination, ideas of sustainable development were easily integrated to create what we call sustainable tourism. The goal of sustainable tourism in this context is to reduce the negative impacts of tourism while enhancing the positive. Sustainable forms of tourism such as ecotourism have been widely embraced as a way to provide local livelihoods and conserve nature; thereby virtually eliminating negative impacts while enhancing positive impacts.

3.3 Sustainable Development and Tourism

Ideas of sustainable development have been highly influential in sustainable tourism. Sustainable development came about as a reaction to the global environmental impacts of human activity and negative environmental outcomes of large development projects. Although sustainability was introduced into the development paradigm in the 1980s, development was (and is) still a top priority, particularly for poverty alleviation. Sustainable development was defined in 1987 at the World Commission on Environment and Development (WCED) as: "Development that meets the needs of the present without comprising the ability of future generations to meet their own needs." (WCED 1987: 42) Implicit in this definition is that development must continue but it must be done in such a way that the resources required are maintained in the long term AKA conservation AKA sustainability.

There is however, a tension between sustainability and development and that tension arises because two different types of systems are at play. First, development is a linear system supposing a starting point and a progression along a path. Economic development for example uses throughputs of matter and energy to produce unending economic growth. This type of development is manifest through industrialization and modernization. Sustainability however, implies a circular or closed loop system. Ideas of sustainability include; recycling, reusing, renewable materials, and mimicking nature in processes. All of these seek to 'close the loops' thus eliminating throughputs of matter and energy. Simply put, nothing is wasted or becomes waste. Everything returns to the system.

How then is the tension between sustainability and development reconciled? In the WCED documents, this is accomplished through the concept of poverty alleviation. The authors note: "A world in which poverty and inequity are endemic will always be prone to ecological and other crises" (WCED 1987: 44). The argument is made that until people are lifted from poverty, they will not be able to care about the environment as their daily needs are more important. As people 'develop' they begin to care for the environment and sustainability becomes important, resulting in reduced environmental degradation. There are two aspects of sustainable development discussed above that are particularly relevant for sustainable tourism.

The first is the tension between the linear system of development and the circular system of sustainability. The second is the assumption that some degradation must occur in the process of development before sustainability can take hold. Sustainable tourism as a form of sustainable development is perhaps the best example of trying to reconcile the linear system of capitalism (as economic development) with the circular or closed loop system of sustainability. Tourism as a capitalist activity relies on the throughput of matter and energy to produce wealth and tourism as sustainability relies on producing a closed loop system. The outcome of this is often that sustainable tourism projects are not sustainable. Martha Honey (1999) provides an excellent example of this with her case of the golden toad in Costa Rica. The Golden Toad's Disappearance from the Monteverde Cloud Forest reserve happened to coincide with the dramatic rise in ecotourism in the area and in all of Costa Rica. Honey (1999) notes that 1500 Golden Toads were found in Monteverde in 1987, by 1989 only one was found and that was the last confirmed sighting of the Golden Toad. The disappearance of the Golden Toad occurred as ecotourism was on the rise in Costa Rica. While it has not been determined what exactly caused the disappearance of the Golden Toad, ecotourism development likely played a role whether it be from a tourist or scientist bringing in a foreign organism or from habitat destruction or water contamination. What is known is that while land was being set aside for conservation of the Golden Toad, tourist numbers in Monteverde climbed from 450 in 1975 to 8000 in 1985 and eventually to over 50,000 by the late 1990s (Honey 1999: 4). What this case illustrates is that although habitat protection and conservation were key components of ecotourism development in Monteverde, unseen and unanticipated impacts from development activity likely caused the Golden Toad's disappearance.

The assumption that some degradation must occur for development is also important for sustainable tourism. It undergirds the very character of tourism in that it (tourism) fundamentally changes the places where it develops. The reaction to these negative impacts has historically been to reduce them to a negligible level. The focus on reducing negative impacts has led to a set of false assumptions about sustainable tourism such as smaller is better and independent tourism (tourists who travel independently rather than in a tour group) is less impacting than mass tourism. The negative impacts of large scale, mass tourism are obvious while those of small scale and independent tourism are less so. Swarbrooke (1999) calls these sacred cows. Sacred cows are value judgements that are treated as fact when there is little or no evidence to support them. He goes on to illustrate using the example of ecotourism. Ecotourism is often seen as a sustainable and small-scale endeavour but Swarbrooke (1999) deftly points out that if ecotourism becomes popular, the scale of activities will grow and the desire for ecotourists to visit new and remote locations will spread ecotourism to more of the world's fragile environments (where it has potential to be as harmful as any other type of tourism). He concludes: "Thus, ecotourists may not be content until they have visited – and brought the mixed blessings of tourism – to every area of the world" (Swarbrooke 1999: 29). The point is that with all types of tourism, impacts occur. Martha Honey's example above also illustrates the complexities of the impacts of tourism. It is not enough to focus on reducing negative impacts when those impacts might be so significantly displaced through space and time that their consequences are felt even before they are identified.

Sustainable tourism is often conceptualized using the three pillars of environmental, economic and social sustainability. Sustainable tourism is said to occur when all three are achieved. Each can be achieved independently though the reduction of negative impacts from tourism and the enhancement of positive impacts. Impacts are monitored through a set of indicators (generally economic, social and environmental). If the indicators are showing few or no negative impacts, then it can be said that sustainable tourism has been achieved. This approach to sustainable tourism is outcome-based. The outcomes of the chosen indicators can point to whether or not tourism is sustainable. Oftentimes, one pillar is preferenced over the other two. In many cases, economic sustainability takes precedence over social and environmental. In other cases, the social or environmental take precedence over the economic. Given that sustainable tourism is an economic activity, it makes sense that economic sustainability would be prioritized over social and environmental. The argument is that without economic viability, sustainable tourism cannot exist and social and environmental sustainability then become irrelevant. This leads to a major tension within sustainable tourism and a crucial question. The tension is between sustainable tourism as an economic (capitalist) activity that necessarily causes negative environmental and social impacts and the need for sustainable tourism to maintain environmental and social sustainability to be economically viable in the long-term. This leads to the crucial question: What is to be sustained?

3.4 What Is To Be Sustained?

In addressing the question of what is to be sustained, we revisit the example of Whitefish from Chap.1. Whitefish, MT has seen a number of resource-based economic activities come and go, with tourism being the latest. Whitefish has benefitted greatly from the presence of Glacier National Park nearby and from the well-established local ski resort. This dramatic landscape has been attracting tourists, second home owners and amenity migrants for more than three decades.

The economic activity generated by tourism has allowed the town to grow and develop, providing more opportunities for visitors and better services for residents. If we view tourism as a form of development for Whitefish, like logging and aluminium processing, what we see is that Whitefish has been resilient in that it is maintaining a trajectory of development. What is being sustained in Whitefish is development. Over time, Whitefish has been able to survive because of its residents' ability to transition from one declining economic activity to another, more profitable activity. Along with these transitions, come changes in land use. National Forest lands that used to be logged are now used for tourism and recreation activities. While it is clear that development has been sustained in Whitefish for more than a century, what is less clear is whether or not Whitefish is an example of sustainable tourism. We know that tourism has been sustained in the region for over 100 years (at ever increasing numbers) with the founding of Glacier National Park in 1910. This growth has caused problems for the environment such as eutrophication and invasive species in Whitefish Lake, urban sprawl and increased impervious surfaces in the surrounding area, causing water runoff and inevitable pollution.

We must also note that although the Blackfeet Tribe has a documented presence in the area for almost 11,000 years, the Blackfeet have largely been excluded from tourism and continue to be marginalized. What this illustrates is that although development has been sustained, it has been uneven, leaving some people and landscapes benefitting and others bearing the costs. So- What is to be sustained in Whitefish? Most people would say tourism is to be sustained. But for what reasons? Again, most people would say that tourism should be sustained because it provides jobs.

This however is a dangerous and superficial perspective. We can look to the ideas of Karl Polanyi for an explanation of why this is so. Polanyi argued that the domination of the market subjects society to its laws and logic (Polanyi 1944). Polanyi theorized that (1) in a market economy everything becomes commodified including land and labor and (2) there is a pervasive belief that markets must go unimpeded. From these, Polanyi made the following arguments: Humans and nature are not produced for sale and therefore are not commodities. Polanyi labeled these as 'fictitious commodities' and said that their exploitation would lead to eventual annihilation of human civilization. Second, Polanyi argued that the domination of the market subjects society to its laws and logic (Polanyi 1944). The outcome of this is that social relationships become part of the economic system rather than the economic system being subsumed by the social (Maertens 2008). States are also implicated in this set of relations as first a regulator and then enforcer of markets (after losing regulatory functions). Polanyi explained these relations in his 'double movement' theorem whereby the transformation of land, labor and money into 'fictitious commodities' endangers nature, people and business. This leads to grievances and resistance with demands for protection. States respond through regulation of the market to prevent the destruction of society itself (Palacios 2002). This is a very important point because when the economic system is the overarching system that the social and environmental systems are contained within, then economics always wins out and social and environmental systems suffer. Does this always have to be the case? Is it possible that what is to be sustained could be answered differently? Next we explore a case study from the Nanda Devi Biosphere Reserve in India to answer that question.

The case of Nanda Devi has many similarities to the case of Whitefish and some distinct differences. Nanda Devi Biosphere Reserve (NDBR) is named for the sacred peak of Nanda Devi (7817 m) and is considered one of the last great wilderness areas of the Himalaya. The area has seen tourism for nearly a century along with other forms of economic activity coming and going, forming boom and bust cycles but keeping the local villages viable. The peak of Nanda Devi is sacred to the local population and was one of the most popular in the Himalaya while it was open to foreign expeditions from 1974 to 1982. However, the uncontrolled nature of mountaineering in the region produced significant environmental degradation along the approach to and at the basecamp of Nanda Devi so in 1982, the region was declared a National Park and promptly closed to all people, tourists and locals alike. The area became a Biosphere Reserve in 1988 and the National Park became part of the core zone of the Reserve while the area surrounding (including many villages) became the buffer zone of the Reserve. In 1992, the NDBR became a World heritage site for its unique biodiversity.

The closure of the park meant that tourism all but ceased in the area, local grazing grounds became off limits, and severe restrictions were placed on the collection of non-timber forest products. Local people were left in a bust cycle, trying to adjust to the lack of economic opportunities. In 2001, an official scientific expedition was launched to determine if the area had recovered enough to be reopened to tourism. The recommendations of the expedition were that the area be opened to tourism but with limited involvement from the locals (as guides and porters) (Kapadia 2001). The expedition itself was controversial and the report spurred action by the villages in the Biosphere Reserve to develop their own set of principles for the development of tourism in Nanda Devi. What is interesting about this case and what makes it different from Whitefish is that the grassroots movement to develop ecotourism in Nanda Devi actually resembled a movement for environmental justice (Bosak 2010). The principles developed by the community focused on equity, sustainable livelihoods, preferencing the marginalized, and promoting what was termed 'biocultural diversity.'

The case of the NDBR illustrates that the answer to the question: What is to be sustained? In this case, it does not necessitate the answer: Development. In the NDBR, communities were trying to sustain the relationship between themselves and the local environment (nature) in order to sustain their culture. Ecotourism was viewed as a means by which local people could accomplish the dual goal of cultural and environmental preservation (noting that their culture could not be maintained without a healthy and functioning ecosystem). Secondary to the goals of cultural and environmental preservation was income generation. If we revisit the ideas of Polanyi, what we see in this case is that the economic system has been subsumed by what I will term the socio-ecological system. Thus, the main concern is for maintaining culture and environment through ecotourism.

How then are these two perspectives different? In most cases like Whitefish, sustainable tourism is an end in itself, another economic activity that will sustain development and economic growth with the consequences to social and environmental systems being secondary. In the case of NDBR, concerns for healthy social and environmental systems are primary and ecotourism is viewed as a means by which those systems can be sustained. In the case of the NDBR, the communities of the Reserve got together, discussed their core values, and decided how ecotourism could be developed to support those values. The principles outlined in the ecotourism and biodiversity conservation declaration guide decision-making for ecotourism in NDBR and represent a process-based, rather than an outcome-based approach to sustainable tourism. What these contrasting but similar cases show us is that sustainable tourism can be re-framed.

3.5 Re-framing Sustainable Tourism

More than 25 years after its inception, sustainable tourism is still struggling with fundamental questions of what is to be sustained and how? What we have seen is that in many cases, what is to be sustained is development, often in the form of economic growth that takes precedence over social and environmental systems. Thus, sustainable tourism can rarely if ever be called truly sustainable. This stems in part from tourism being a vehicle for global capitalism and as such, the economic system subsumes the social and environmental systems. When this happens, the economic system carries out its purpose of unending growth and transformation of resources into economic value. The danger is that economic decisions take precedence over social and environmental and the latter suffer at the expense of the former. The example from NDBR illustrates that this does not have to be the case. It is still possible to preference the social and environmental over the economic. Doing this however takes a different approach, one that we can learn from in our attempt to re-frame sustainable tourism. There are two components to re-framing sustainable tourism that have been explored thus far in this chapter, and that will be expanded upon to form a pathway to a new perspective. These are: systems thinking and ideas of resilience.

In the case of the NDBR, the opening of mountaineering on Nanda Devi in an uncontrolled manner produced certain feedbacks in the socio-ecological system as did the closure of Nanda Devi to all people in 1982. Tourism from mountaineering in the period of 1974–1982 produced negative environmental impacts particularly with regard to cutting of trees for fuel wood, disturbance and hunting of wildlife, and human waste and pollution. In turn, the closure of the NDBR in 1982 after reports of environmental degradation had its own feedbacks for local communities and the environment. Villagers had to sell or slaughter their sheep and goats because traditional grazing grounds were declared off limits. This meant that people had less money, meat and wool. The closure of the grazing grounds also meant that certain sacred locations became off limits and people could not perform religious rituals based in the worship of the peak of Nanda Devi. What this illustrates is that separating the economic, social and environmental often leads to unwanted results. When tourism in the form of mountaineering was allowed to proceed uncontrolled, the result was that the environment became degraded within a few short years. When the environmental aspect was preferenced heavily in the form of the closure of the area and the creation of the Biosphere Reserve, the social and economic resilience of local people suffered.

If we regard social and ecological systems as linked, dynamic, complex and adaptive then we can see that society and environment are closely linked and that artificial separation of the components can lead to unwanted and unanticipated feedbacks. The socio-ecological systems approach has a focus on resilience and as Folke (2006: 254) points out: "It is argued that managing for resilience enhances the likelihood of sustaining desirable pathways for development in changing environments where the future is unpredictable and surprise is likely."

Resilience is often conceptualized through the example of a ball in a shallow basin (Walker and Salt 2006). The basin represents the normal range of variability in a socio-ecological system and as long as the ball (representing the state of the system) remains within the basin, the condition of the system is described as 'normal.' However, because systems are ever-changing, perturbations will occur that will cause the metaphorical basin to tilt or be reshaped. If perturbations are large enough, they will cause the ball to roll out of the shallow basin. This represents the transformation of the system into something different (another system). Resilience therefore is described as the ability of the system to retain its integrity (the ball in the basin) and maintain its trajectory of development after it has been disturbed. Folke et al. (2010) citing Walker et al. (2004: 4) define resilience as: "The capacity of a system to absorb disturbance and reorganize while undergoing change so as to still retain essentially the same function, structure and feedbacks, and therefore identity, that is, the capacity to change in order to maintain the same identity." What this definition illustrates is that socio-ecological systems are dynamic and complex, constantly adapting and renewing. What is key is the preservation of a system's ability to adapt to changing conditions while still maintaining its integrity. What resilience thinking teaches us is that socio-ecological systems are not static and simple as they are often conceptualized in sustainability.

In the context of sustainable tourism as a form of development, resilience is particularly salient. The environments that sustainable tourism depends upon for its existence are constantly changing. In addition, many of these changes are beyond control (at least at the local level) and can include global climate change, changing market demands, natural disasters, and political instability. The effects of these changes are often unanticipated and require the capacity to adapt in order to maintain resilience.

However, this leads to the question what type of resilience and for whom? What the example of Whitefish, MT illustrated is that economic resilience can be maintained in the long term but often if not always comes with social and ecological degradation. Therefore, the answer to this question, I argue, is that the resilience of socio-ecological systems is what 'should' be maintained, particularly over economic resilience. We see evidence of the effort to link and maintain socio-ecological systems in the Nanda Devi Biodiversity and Ecotourism Declaration, whereby the term bio-cultural diversity is used to show that cultural and biological diversity are intimately linked. In addition, there is a focus on equity and preferencing the marginalized for economic opportunities, preventing exploitation, and revering nature above all else that clearly de-prioritize the economic over the social and ecological.

On the surface, this presents a problem for tourism in that it is inherently an economic activity and is a driver of global capitalism. However, the problem is only superficial if we make the argument that sustaining healthy socio-ecological systems will then lead to healthy economies. This is opposed to the argument that healthy economies lead to healthy social and economic systems that we often see forwarded through sustainable development rhetoric. It is this shift in perspective that allows us to re-frame sustainable tourism.

3.6 Tools

Adaptive governance approaches have been identified as a means by which resilience of socio-ecological systems can be enhanced (Folke et al. 2005). Walker et al. (2002) proposed a four-step model for participatory resilience management that could serve as an important tool for adaptive governance of sustainable tourism. This model first involves understanding what the system is composed of, its historical context, and the boundaries and linkages of the system. Once these are identified, the second step is to understand what the system needs to be resilient. This involves scenario building and visioning. The goal of the second step is to understand the range of outcomes and to select the most desirable. This is of course a value-laden process but one that should be informed by as much information as possible. Step-3 involves modelling system dynamics to understand how variables interact to affect resilience. In this case, modelling can be anything that allows people to understand the abstraction of the system. The final step involves evaluating the entire process, understanding the critical variables for resilience, and developing actions that will enhance resilience. These often manifest in the form of policies. As Walker et al. (2002) explain: "the policies are aimed at a set of rules (incentives and disincentives) that enhance the system's ability to reorganize and move within some configuration of acceptable states, without knowing or caring which particular path the system might follow." This is all based on the idea that a common understanding and policies can be developed through consensus. Folke et al. (2005) note that this often occurs through self -organization and social networks and can be enhanced through what they term as 'bridging organizations' such as NGOs that can help to create the conditions for adaptive governance to be effective.

In the case of the NDBR, having the realization that the core zone might be reopened to tourism thereby drastically changing the socio-ecological system brought about the self-organization of a workshop where a set of principles were developed. While the path to ecotourism was unclear, the set of principles in the Nanda Devi Biodiversity and Ecotourism Declaration set parameters for acceptable states of the socio-ecological system and began to create the conditions for adaptive governance. Next, several bridging organizations became involved. These were largely NGOs both at the local and international level who helped locals to scale up by networking them with academics and activists. The outcome of this is the development of a community-owned ecotourism enterprise that follows the principles outlined by the Nanda Devi Biodiversity Conservation and Ecotourism Declaration. The missing component in this case has been policy formulation and governmental actions that support adaptive governance. Instead, the government in general and the Forest Department (in particular) are seen as adversaries and their policies are seen as detrimental to maintaining resilient socio-ecological systems. What this illustrates is that in order for adaptive governance to be an effective tool to promote sustainable tourism (as a resilient socio-ecological system), it must occur at all scales and government policies must be developed to support creative pathways to resilient systems.

References

Bosak, K. 2010. Ecotourism as environmental justice? The case of the Nanda Devi Biosphere Reserve. *Journal of Environmental Philosophy* 7(2): 49–74.

Britton, S. 1982. The political economy of tourism in the third world. *Annals of Tourism Research* 9(3): 331–358.

Fletcher, R. 2011. Sustaining tourism, sustaining capitalism? The tourism industry's role in global capitalist expansion. *Tourism Geographies* 13(3): 443–461.

Folke, C. 2006. Resilience: The emergence of a perspective for social–ecological systems analyses. *Global Environmental Change* 16: 253–267.

Folke, C., T. Hahn, P. Olsson, and J. Norberg. 2005. Adaptive governance of social-ecological systems. *Annual Review of Environmental and Resources* 30: 441–473.

Folke, C., S.R. Carpenter, B. Walker, M. Scheffer, T. Chapin, and J. Rockström. 2010. Resilience thinking: Integrating resilience, adaptability and transformability. *Ecology and Society* 15(4): 20.

Hipwell, W. 2007. Taiwan aboriginal ecotourism: Tanayiku Natural Ecology Park. *Annals of Tourism Research* 34(4): 876–897.

Honey, M. 1999. *Ecotourism and sustainable development: Who owns paradise?* Washington, DC: Island Press.

Kapadia, H. 2001. *Report and recommendations of the Nanda Devi Sanctuary Expedition 2001*. Submitted to: Shri N.N. Vohra, President. New Delhi: Indian Mountaineering Foundation.

Maertens, E. 2008. Polanyi's double movement: A critical reappraisal. *Social Thought and Research* 29: 129–153.

Palacios, J. 2002. Globalisation's double movement: Societal responses to market expansion in the 21st century. In *Proceedings of the eighth annual Karl Polanyi international conference "Economy and Democracy."* Mexico City.

Polanyi, K. 1944. *The great transformation*. New York: Rinehart & Company.

Schellhorn, M. 2010. Development for whom? Social justice and the business of ecotourism. *Journal of Sustainable Tourism* 18(1): 115–135.

Swarbrooke, J. 1999. *Sustainable tourism management*. Wallingford: CABI.

Walker, B., and D. Salt. 2006. *Resilience thinking: Sustaining ecosystems and people in a changing world*. Washington, DC: Island Press.

Walker, B., S. Carpenter, J. Anderies, N. Abel, G.S. Cumming, M. Janssen, L. Lebel, J. Norberg, G.D. Peterson, and R. Pritchard. 2002. Resilience management in social-ecological systems: A working hypothesis for a participatory approach. *Conservation Ecology* 6(1): 14.
Walker, B., C.S. Holling, S.R. Carpenter, and A. Kinzig. 2004. Resilience, adaptability and transformability in social–ecological systems. *Ecology and Society* 9(2): 5.
Weaver, D. 2001. *The encyclopedia of ecotourism*. Wallingford: CABI.
World Commission on Environment and Development (WCED). 1987. *Our common future*. Oxford: Oxford University Press.

Part II
Frameworks

Chapter 4
Frameworks for Tourism as a Development Strategy

Art Pedersen

Abstract The chapter addresses issues in using tourism for wide development purpose, supporting economic and also social and cultural objectives. It suggests changes in working approaches at the international and community levels to do this. The approaches recommend processes for creating practical entry points to bring diverse interests together to reflect upon, identify, and prioritize the range of tourism's potential, helping a community to become more vibrant and resilient. The chapter mentions the challenges in operationalizing the concept of sustainable tourism. It suggests focusing on tourism's purpose; asking questions about what resources tourism would bring to a community or region, presents a practical option for those who deal with the push and pull of tourism development and conservation. It outlines some of the considerations of community engagement and refers students to more detailed resources.

Keywords Community development • Community tourism • Development strategy • Community engagement • Tourism integration

4.1 Introduction

McCool (1999) writes, "Perhaps the most fundamental challenge confronting the tourism industry today is its ability to contribute to three fundamental goals of human welfare; (1) providing for economic opportunity; (2) enhancing quality of life; and (3) protecting our cultural and natural heritage." He goes on to state that, "tourism and recreation can be an agent for societal development, a tool deliberately and carefully chosen to address these goals". Rather than viewed separately, tourism can be strategically and proactively integrated into community development strategies, helping a community become more vibrant and resilient (McCool 1999).

A. Pedersen (✉)
Independent Consultant, Helsinki, Finland
e-mail: a.pedersen24@gmail.com

S.F. McCool, K. Bosak (eds.), *Reframing Sustainable Tourism*,
Environmental Challenges and Solutions 2, DOI 10.1007/978-94-017-7209-9_4

Deciding tourism's purpose; asking questions about the resources tourism would bring to a community or region, and using it as a tool contributing to a wider range of needs, is the theme of this chapter. Tourism obviously has a predominantly economic development focus. This is frequently accompanied with statements connoting broader and deeper social purpose; the need to protect a community's sense of place or quality of life; safeguarding those resources that are both tourist attractions and resources associated with a community's pride and cultural values. It may be argued that to the extent a community explores a range of its values, including cultural and historical values, a wider dialog on community needs and desires may be fostered and with this a wider range of tourism benefits generated.

There is logic and precedent for taking this wider approach. Sustainable development is meant to be cross cutting; both economic and social statistics are tracked by most countries; the European Union tracks in detail economic and social indicators like education, sustainable transport, public health, and good governance (European Commission 2013). General definitions of sustainability increasingly use the idea of well-being or quality of life, this broader and deeper focus taking into account the importance of social and cultural capital, such as education and health because, "they cannot be substituted with other capital forms" (Moscardo and Murphy 2014). Recent tourism-focused academic articles support the industry's wider use and integration. Saarinen (2013) advocates a regional development, with tourism having, in addition to economic goals, qualitative goals like quality of life improvement and well-being. He asks, "Who are we responsible to in sustainable tourism development" (Saarinen). Other experts propose that different types of tourist activity are assessed and positively linked to social and cultural elements. "Failure to consider tourism's integration with an array of development options limits the extent which tourism can be seen as sustainable" (Moscardo and Murphy 2014).

Having as a goal enhancing tourism's potential for wider development integration, this chapter offers suggestions for expanded thinking on the part of global organizations charged with guiding sustainable tourism policy. It offers practical processes for creating entry points to bring diverse interests together to trigger or catalyze, in addition to economic, social and cultural considerations. The objective is here to provide opportunities for communities to reflect upon, identify, and prioritize the fuller range of tourism's potential; attempting to make tourism greater than the sum of its parts, going beyond its recognised ability to produce economic benefits thereby contributing to more resilient and vibrant communities, enhancing quality of life, protecting cultural and natural heritage, and along with this, generating economic opportunity.

There is much theory and discussion on sustainability and sustainable tourism; perhaps a dose of frustration is evident in taking this broad, complicated and obtuse concept to practice. But tourism is here to stay and future policy makers and the next generation of community leaders will still need to deal with and implement tourism policies and projects. There are no easy answers, but there may be better ways to work, positively influencing tourism's wider usefulness as a development tool, improving its integration into the types of development strategies that would help support the three pillars suggested by McCool.

4.2 The Varied Uses of Tourism

It seems that wherever one goes, tourism is being used for something other than its intention, the pursuit of some desirable experience away from home; (if one is lucky enough, and depending on one's tastes, perhaps a week relaxing in a tropical paradise).

The pursuit of broadly defined sustainable or responsible tourism, travel in small groups, producing little environmental and cultural impact, may very well serve an entirely different purpose than providing an enjoyable holiday. For those involved with conservation issues, tourism frequently has as a goal to generate an alternative income source for local people living on the borders of protected areas; here the purpose is creating incentives to stop hunting an endangered species or the practice of overfishing or lessening a trade in illegal artifacts.

Tourism may be used by a community for the purpose of halting the deterioration of an important cultural landscape. At Cinque Terre in Italy, the community embarked on a campaign of producing local agricultural products to sell to visitors, providing jobs for young people so they remain in the area and continue to farm that landscape's ancient terraces (Rebanks 2008).

Tourism may be a means to finance protected areas through visitor fees. This is the case with Iguazu Falls in Argentina, where visitor fees at the park also generate significant financing for other Argentine national parks.

Tourism can address geo-political issues such as immigration, providing jobs to those workers in the sending countries who without work would be forced to migrate. During the 1980s the tourism policy activities of the US Federal Commission for the Study of International Migration and Cooperative Economic Development had this focus as one of their goals (Pedersen and Ceballos-Lascurain 1990). The Hercules, EU Project, training young people from countries in North Africa in tourism management, so that they might better manage the heritage resources of their own countries, can be associated with the issue (European Commission, Erasmus Mundus 2014).

Tourism may be linked to a community's way of life and sense of place. This may include education, access to recreational and social amenities, spiritual concerns, and community and family cohesiveness. Tourism prompts new and expanded community facilities to ensure well maintained local areas, entertainment and recreation venues and transport services; without tourism, some communities may not be motivated to improve street maintenance and water services. Tourism can improve community pride, as locals recognise its distinctive characteristics; this increase in pride encourages the celebration of local events and cultural activities (Tourism Queensland 2014).

These effects are backed up by examples. In the US town of Myrtle Beach, South Carolina, tourism revenues helped the expansion of educational facilities (Myrtle Beach 2011).

In Bohol in the Philippines, the Lobac River and the town's church provide a popular tourist venue. Bohol's Tourism Centre and the associated paved roads constructed for tourists, also provide improved access to the town for farmers and other residents. The Centre's free wi-fi enables students to access the internet for their studies and families to connect with relatives working overseas (Coffey International Development 2013).

In Nepal's Solu Khumbu District, many thousands of trekkers come yearly to view the scenery and experience the Khumbu's Sherpa communities. The Himalayan Trust, founded by Sir Edmond Hilary, with support from the tourism industry, implemented a teacher training programme for the schools closest to Mount Everest. Also, in the Khumbu District, the Kunde Hospital which serves over 10,000 Sherpa people would probably never have been supported over the years without mountain tourism and its secondary effects of raising the awareness of international donor agencies (Himalayan Trust 2014).

Suchitoto, a sixteenth century town in El Salvador, largely abandoned during El Salvador's civil war, is characterized by one-story dwellings with red clay roof tiles, linked by arcades and interior courtyards. As Suchitoto was repopulated, new construction impacted the towns' character and aesthetic integrity. With these growing threats, but also a revival of tourism, the town established a school to educate young craftspeople in traditional construction techniques. Young people now find work in the ongoing historic preservation efforts fueled by community pride with impetus from domestic and foreign tourism (World Monuments Fund 2014).

At the Buddhist pilgrimage site of Lumbini, Nepal, the site provides recreation areas for the local Hindu population. At Wadi Rum in southern Jordan, tourism enables a women's cooperative to maintain traditional arts and crafts, arguably adding to community cohesiveness.

The list goes on. The point is that tourism may serve many purposes; it can be a means to the end and contribute to an array of social, cultural and economic benefits.

4.3 Issues for Tourism as a Development Strategy

However, experiencing tourism's effects, such as support for a hospital in Nepal or the sudden access to Wi-Fi in the town of Bohol in the Philippines, differs from proactively, with purpose, using tourism for strategic and wide-ranging goals. Purpose is seen in a number of conservation and pro-poor tourism efforts concerned with providing an alternative or supplementary income source to disadvantaged people. In project work, organizations such as the UNESCO World Heritage Centre, Rare, and The Nature Conservancy have used tourism in this fashion, specifically to aid biodiversity conservation at a number of World Heritage sites. International agencies have used tourism value chain analysis and developed projects in the

Mekong area of Vietnam and Laos, specifically to maximize economic benefits for local people. There are many examples of the economic effects of tourism on communities as well as tourism-conservation linkages (Mekong Tourism Coordinating Office 2009).

What is more difficult to find, are examples of how communities have strategically developed tourism, tying it to the broad range of socio-economic benefits that it is capable of delivering. For instance, it is hard to find examples where communities have specifically targeted tourism to generate economic benefits as well as also build community pride. While one can see potential in linking tourism to transport, education and health planning, information on proactively generating a suite of benefits from these sectors is scarce. To be sure, tourism has been the impetus for transport construction, as a means to get visitors to attractions, look at the Olympic Games or airports built for charter flights in developing countries; but it's difficult to find examples of where communities have proactively carried out initiatives such as building a resort but also planning the access roads to and from the tourist attraction so that they might serve diverse local community needs; farm to market roads benefiting local agriculture, or improved community access to education and health care. Or on the education and health issues, it's hard to find examples such as communities and the tourism industry designing the information produced for visitors on local and neighboring cultural attractions so it might also be used in schools to educate local students. Or planning health services that provide inoculations against a local pest bourn ailment so that the service could also be offered to tourists, who may need this protection during their visit, and who may also help to supplement financing of the service for local citizens. There are numerous segments of community life where tourism may be slotted in to aid desired community goals and objectives. But with this potential, why are examples so difficult to locate of communities combining and maximizing the entire range of potential benefits through a suite of activities; why is tourism's potential for its wider use so underutilized? What might be some of the inhibiting issues? (see Fig. 4.1).

4.4 Connecting the Dots

One key challenge, for using tourism for wider purpose seems the need for greater understanding and expanded thinking by decision-makers, the different constituencies, about how tourism may be linked more holistically to community socio-economic concerns. There seems the need for venues and practical initiatives that provoke this wider thinking and discussion. Several examples help to offer insight into the difficulties in making the necessary linkages.

In a study carried out in the State of Montana, tourism professionals were surveyed and asked to provide their ideas on the desired conditions that tourism could generate or should sustain. These professionals responded that one condition should be that tourism would enhance Montana's quality of life. However, when asked about the most important tourism indicators for monitoring desired conditions, such

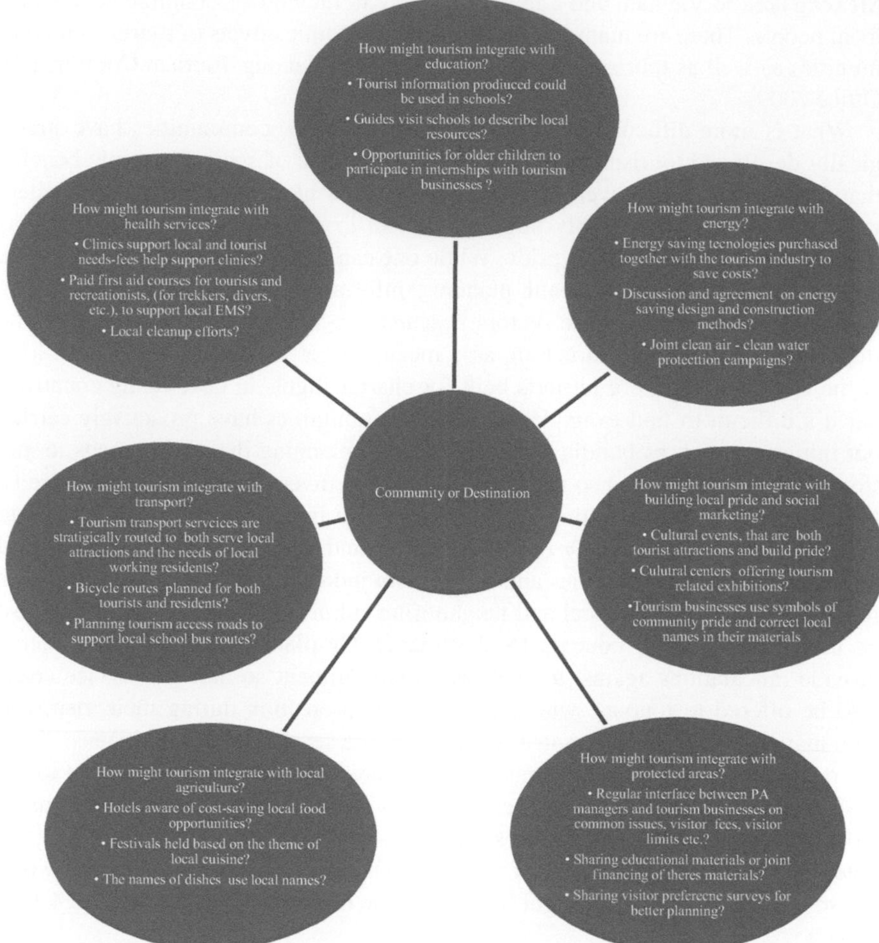

Fig. 4.1 Sample questions for opening discussions on potential tourism-community linkages

as quality of life, they reported increasing hotel bed occupancy rates as a one such measure of success (McCool 1999). While hotel occupancy rate can indicate economic success, its connection as an indicator of an increasing quality of life is less clear. In this case, there appears a need for more clarity between the desired conditions of the professionals and their awareness of how tourism might best serve to meet the desired goals.

Another type of disconnect was found in the results of a tourism integration study on a number of Transition Towns in the United Kingdom (Waddilove and Goodwin 2010). Transition Towns, a movement founded in the United Kingdom and now spread to over 3000 communities around the world, aims to create resilient communities that have the capacity to respond to the challenges of energy dependency, climate change and global economic instability. Grounded in community facilitation and group process, a Transition Town strives to become independent of external energy and food sources and maintains commerce through community trading mechanisms. Working groups on local food, energy, waste, economics, health, education, transportation, arts, housing and local government liaison are used to create spaces for conversations around topics of sustainability and local resilience.

What was found, in the aforementioned study of tourism's integration, was that while it was generally agreed that tourism has to be addressed as part of efforts to increase the community's sustainability and resiliency, communities are unsure of how to go about incorporating tourism into the broader community fabric of their different working groups; it seems more direction is needed for exploring tourism's use in the Transition Town model.

A UNESCO World Heritage Decision-Making Framework was developed and used for the aforementioned Palau Workshop. The workshop's goal, using a decision-making framework being developed by UNESCO World Heritage Centre, was to begin to determine a site's purpose for tourism, what tourism would sustain, and the conditions and experiences that should be maintained and promoted at the site. In the case of Palau, what was being explored was the cultural tourism offer at the mixed (culture and nature) Rock Island Southern Lagoon World Heritage site, deciding on its purpose and integrating it into the current tourism offer geared mostly to marine tourism, particularly diving. The exercise was also intended to provoke wider thinking on a country tourism strategy that helps Palau communities to become more vibrant and resilient.

What was observed was that while participants, both from the tourism industry and the public sector, focused easily on tourism's economic purpose and benefits, they were much more challenged in seeing the full range of potential cultural connections and advantages, even though the site's cultural assets could benefit the existing tourism offer, for example, as a way to help relieve pressure on congested dive sites. It seemed that what was needed was time and information for creative strategic discussion to explore tourism's more unfamiliar links to cultural purpose (see Fig. 4.2).

Perhaps making the links to tourism's broader use should be the expected reality. Experiences with World Heritage properties have consistently shown the different realities and dynamics under which key constituencies operate. For example, managers of a protected area or a cultural asset such as a museum may not have full knowledge of the concerns of the tourism industry, and the tourism industry may not be familiar with the issues of managing natural and cultural resources. This is in addition to the realities of other frequent implementation constraints; project proposals designed with only an economic emphasis, top down planning by public agencies, the capacity needs of local communities to better enable participation in bottom-up planning, short project timeframes and unsustainable funding mechanisms for project follow-up, among others.

Workshop Opening - "Take a minute to think about, and write down, what you think the purpose of tourism should be, (sustain local economies, protected areas, traditions, etc.) and why".

"What are the World Heritage site's resource attributes"? (An example of a site resource attribute included, the site's archaeology resources or traditional cultural resources and practices such as dance, or food, etc.)

"What are the World Heritage site's attractions and related products or services"? (Related to resource attributes, attractions may be a local foodstuff or a tradition of a local dance; products the development of dance related festivals and local food products to sell in hotels.)

"Thinking back on the opening question, what does tourism sustain, what do we wish tourism to accomplish with each attraction and tourism product"? (Examples included, improved economic livelihoods, jobs for young people, helping to finance conservation, building pride in traditional cultural practices, financing protected areas or schools)

What are the specific purpose(s) of emphasizing a specific resource attribute, a related the attraction and its related products? (Example, archaeology and perhaps the development or strengthening of a local museum or exhibition and exhibiting Palauan pottery. Purpose, to provide employment for senior citizens of a specific community.)

Additional questions:

"What should the purpose of tourism be at the World Heritage site"?

"What are the experiences you wish to offer visitors with the above tourism offer"? For example, what type of museum will it be and what experiences will it provide visitors?"

"World Heritage site is unique, special and globally important, what are the experiences that should be offered to visitors to reinforce or confirm that uniqueness"?

"What is needed to develop the attraction and its related products?" (For example, identification of the tourism markets that serve to produce the desired results or reaching consensus in a particular community.)

"How are these cultural products best integrated into the existing tourism structure?"

"Is the infrastructure generated by these markets compatible with the presentation of the site's Outstanding Universal Value (The term used in World Heritage that connotes the significant value making it eligible for World Heritage status), sense of place and local values?"

"How might World Heritage be used to help accomplish the desired purpose of tourism?"

Fig. 4.2 The series of Palau workshop questions

4.5 Ideas for Broader Tourism Integration

So, what might be some of the specific actions that could be taken to help facilitate the integration of tourism into a broader development strategy, taking advantage of this greater potential? Expanded thinking from the global to the local level would most certainly benefit tourism's potential for broader integration into community development strategy thinking and implementation efforts.

4.5.1 *Sustainable Tourism and the Question of Purpose*

One change of thinking at the global sustainable tourism policy level, that would have positive repercussions at the local level, would be a shift of focus of the sustainable tourism question. Currently, the sustainable tourism focus is oriented toward making the tourism industry more environmentally and socially conscious, in particular with energy efficiency and generating local economic benefits without negatively impacting local cultures.

For those working on the practical initiatives of tourism's integration into development, global sustainable tourism policies, generated by international agencies and organizations could aid integration if they focused less on the question of how to sustain tourism and emphasized more, the issue of what tourism should sustain; its purpose. Asking constituencies to ask themselves why they are developing tourism, what should tourism sustain, could generate a number of far reaching benefits.

For one it would help people to better grapple with the many decision-making dilemmas raised by tourism development. At a destination, should a planned tourism access road benefiting one larger community be built if it negatively impacts the agricultural land of a smaller but important rural community whose products may be sold to the tourism industry? The question of what tourism should sustain is important here. Perhaps supporting the smaller agricultural communities is more important because their potential instability, if they don't receive benefits, may negatively impact security and ultimately the regional tourism industry. Tourism questions of this kind require a healthy dose of local thought and response on what is important to the community and region. Reflecting on the question of tourism's purpose helps provoke thinking on a range of tourism ideas related to community needs and values, the motivations related to the different constituencies. It enables analysis of potential actions for a wide range of interest groups.

Identifying tourism's purpose is also important from an operational standpoint. Where sustainable tourism has a purpose, it is palpable, not an abstract concept, and provides a direct pathway for substantive community discussion. It becomes a tool or a vehicle to get somewhere. When something is considered a tool, it inspires an element of control shifting the emphasis onto the community's or host's management responsibility, thereby empowering them.

The question of purpose also allows constituencies to focus in on the details of implementation, what each one can do to make the desired goals a reality and avoid or minimize pitfalls. This makes it easier to track, to determine if it's doing its job, fulfilling its purpose or not. Also, tourism can be controversial for constituencies dealing with the push and pull of conservation and development. In defining an agreed upon purpose, tourism is presented on more neutral ground, it is not good or bad; its utility is based on how one uses it.

In addition, at the international level, asking the question of "what should tourism sustain" could help reformulate project development policy at the agencies in question. This could influence project funding initiatives to consider projects whose focus question is not what resources are available in a destination for tourism, but what resources tourism would bring to the region to meet resident and non-tourism business needs.

4.5.2 Criteria, Principles and Sustainable Tourism Charters

Global sustainable tourism policies would be enhanced by providing guidance to communities and destinations on the management and development frameworks with processes that encourage bringing diverse interests together for defining tourism's purpose. For example, the Global Sustainable Tourism Council (GSTC) which serves as the international body for fostering increased knowledge and understanding of sustainable tourism practices recommends that destinations have "a visitor management system for attraction sites that includes measures to preserve, protect, and enhance natural and cultural assets" (GSTC 2013). The International Council on Monuments and Sites (ICOMOS), in its International Cultural Tourism Charter states that, "Before heritage places are promoted or developed for increased tourism, management plans should assess the natural and cultural values of the resource" (ICOMOS 1999).

In general, management plans suffer from low rates of implementation and quick obsolescence. They are usually written by an external consultant and there can be an issue of local buy-in. But management plans can be very useful tools if used as platforms to bring constituencies together to discuss what tourism should sustain; working out ways and means of using tourism to meet desired development outcomes.

If these aforementioned recommendations on planning could also offer brief complementary information on better ways to work, perhaps in an annex so that criteria and charters would not have to be rewritten, it would help to guide the integration work on the ground. For instance, a supplemental paragraph in an annex advocating a process whereby the associated community constituencies produce the plan, developing clarity of purpose in their goals and objectives and market preferences and learning along the way. This would augment the useful guidance these documents provide.

Planning is only one example. There are a number of robust tourism development and management decision-making frameworks that have proven their utility in bringing constituencies together. The Recreation Opportunity Spectrum (ROS), for example, used mainly in natural settings, is now also being used at historic centres, historic urban and cultural landscapes. It addresses the question of appropriate visitor experiences and conditions, all issues related to who will use the resource and for what purpose, along with issues such as visitor limits to maintain desired experiences.

If the criteria-type documents, produced by the appropriate international organizations, contained a package of these frameworks with processes that help engage constituencies to reach consensus on core values and motivations, it could further generate positive results in policy changes for funding community and regional development project efforts. This could result in more robust and focused destination training programmes, focusing on better ways to work; something that is very much needed.

4.5.3 *A Bank of Information*

If the question of "what should tourism sustain" is widely adopted by organizations, a bank of practical information on how communities have been able to organize themselves to ask the question of tourism's purpose and how this has been linked to community needs would be a global benefit for communities and destinations. While it would be important to begin to collect these examples within the appropriate global organizational networks it would also be important to link the information to regional outlets where shared language and experience could facilitate its usefulness, for example, in the Balkans or in Central America. Countries such as China would be well served to have their own banks of information. The information might be organized at prominent regional universities or institutes, for example like the Centro Agronómico Tropical de Investigación y Enseñanza (CATIE) in Costa Rica, who could then be able to organize needed training efforts for the destinations in their particular zones of influence.

4.6 Initiatives to Promote Tourism's Wider Integration

At the destination level tourism's integration into development suggests the need for venues to discuss the linkages for integrating tourism into a broad development strategy. Besides such processes as management planning, what might be several preferred initiatives; vehicles, to set up situations for asking the question of tourism's purpose so as to provoke wider strategy thinking? What initiatives might have potential to resonate with important constituencies without having to reinvent the wheel?

4.6.1 *Existing Community Groups and a Tourism Liaison*

A practical solution offered from the Transition Towns study was to simply integrate tourism into the existing community working groups, those groups dealing with transport, education, health services. etc., rather than forming a separate tourism group. The study mentioned that this idea was conceived as facilitating a more holistic approach to tourism development where tourism's linkages could be seen and made to the other activities (Waddilove and Goodwin 2010).

This is reasonable, but related to this it is well known that there are intrinsic barriers between key constituencies. Experience has shown, certainly at many World Heritage sites, the utility of someone on the staff of a regional authority or within a protected area, who makes the link between the protected area, the community and the tourism industry. For example, at the Petra Development and Tourism Region Authority in Jorden, it was recommended that the predominantly economically focused authority would benefit from a staff member who was familiar with the range of issues of the archaeological site and its conservation, the community and the tourism industry. It could be suggested that this type of professional, may, with Twenty First Century sustainability issues in mind, provide a key link in facilitating and informing those different working groups mentioned in the Transition Town model, better enabling tourism to be integrated within the wider focus.

4.6.2 The Use of the Interplay Between Cultural and Economic Interests

It can be argued, that to the extent a community explores and then incorporates its cultural and historic values into its tourism decision-making strategy, a wider dialog on community needs and desires may be fostered, beyond solely the economic. Activities where cultural values and considerations come into play can complement the more usual solely economic-focused interests. But it is also clear that economic interests are generally the primary drivers and these interests can hold the greatest sway over the final outcomes. Initiatives that have the ability to advance the social and the cultural and historic but also have ties to the business community and economic constituencies would seem to stand the best possibility of success in exploring the full range of tourism potential.

Several approaches that tie cultural and historic considerations to economic interests that may enable communities to reflect upon, identify, and prioritize the full range of tourism's potential are suggested in the following paragraphs.

4.6.2.1 Cultural Events

A strategic use of cultural events could help to trigger a look at tourism's larger purpose further enabling constituencies to explore the question of what is tourism to sustain. Research has shown that it's possible for public sector marketing efforts to utilize events strategically to bring long-term economic and social benefits to the location and its community. Events can become an integral part of the collective communities' psyche in terms of celebration of culture and as a demonstration of civic pride providing social as well as economic benefits. The social benefits accrued are likely to include, yes, enhanced civic pride, and also higher levels of community involvement (Pugh and Wood 2004). With cultural events, it could also be thought that these benefits, if strategically planned, would be able to be used to explore other spin off activities, such activities as enhanced educational programmes for schools and adult learners adding to their utility of purpose.

One of the ideas ultimately generated with the aforementioned Palau initiative was an idea to reintroduce the idea of a Taro Festival to the islands. The Festival had been a one-time event and had taken place several years ago as a fair promoting local culture and pride. Taro, a food staple in the Pacific islands, was consistently listed as a cultural icon that had to be recognized in a tourism product. Eventually, there was a look at tourism's larger purpose and discussion of the usefulness of the festival as a vehicle for generating cultural pride but also for promoting marine conservation issues; grown in coastal wetlands the taro crop cover can aid the protection of the coral reefs at the World Heritage site. In tandem, this triggered a discussion on the use of taro products in a variety of foodstuffs that could be developed as commercial products. Interestingly, the introduction of this festival theme also introduced the issue of a Palau cultural copyright law now being debated by government officials.

4.6.2.2 Mapping Heritage Assets

If a destination, its associated protected areas and supporting organizations and businesses, have an interest in heritage tourism-led development, an initiative that could help to facilitate a wider dialog on community needs and desires could be the process of identifying, prioritizing, and mapping regional cultural and natural heritage tourism-related assets. This initiative, using a region's or community's tangible and intangible heritage assets, brings together diverse interests in a region-wide reflection to define their goals and objectives and help to see where practical linkages and overlaps might be encouraged. Experiences in the Douro Valley of Portugal and several areas in the US, Mexico and South America using the National Geographic Map Guide initiative have shown the process as a useful one to make connections to a variety of constituencies. In these cases, the initiatives formed stewardship councils generally using existing tourism associations and protected area management and then through surveys identified and prioritised community-based heritage attractions and their related products. One of the hooks for businesses is the marketing benefits of being associated with a particular attraction identified and described on the National Geographic paper and digital Map Guide. One of the hooks for protected areas is the distribution of visitation – if protected areas are well-managed for visitation. Aside from the economic, the community interest is linked to the element of building community pride in the tangible and intangible heritage assets identified by community members.

While the National Geographic experiences are mentioned, a community or region could organize similar initiatives and promote the heritage assets using their own paper or digitized map and website. An objective of the initiative might also have a small percentage of the profits from those businesses, associated with the heritage assets on the map, going back into a fund to aid on-going efforts the community deems appropriate.

4.6.2.3 Interpreting Heritage

Another activity that links culture and economy, and could be used as an initiative and platform to discuss tourism's integration into development, is an interpretation initiative of important heritage assets. On generating community investment, heritage funders suggest that they are more willing to invest in projects if there is a strategy that spells out how different sites or heritage assets can complement each other to create storylines that lure visitors across the landscape, creating with it a greater potential market (Rebanks 2010). Creating these shared story lines or interpretation messages associated with the various heritage assets could be a complementary initiative that links different interests. If these interpretation messages are linked to related local products and/or attractions, the situation can provoke thinking on such broad issues as transportation connections for visitors and local residents.

4.6.2.4 A UNESCO World Heritage Decision-Making Framework

Examined through the lens of a World Heritage sites' Outstanding Universal Value, World Heritage is well-suited as a point of reference for exploring the tourism issue and generating needed debate. The World Heritage inscription process, a 2-year long process, or the desire to define a heritage tourism offer after inscription, are attractive themes to engage a wide range of diversified social, environmental and economic interests. This can help identify and clarify the points of agreement and conflict for reaching the consensus needed to help refine both tourism management and development paths. In addition to its potential as a vehicle for generating public and private sector input; World Heritage status if used proactively, can help catalyse a number of actions. These include, enhanced image and place making, raising public awareness, better planning, the potential for fundraising to meet unmet conservation costs, all may benefit from having World Heritage status.

A UNESCO World Heritage Decision-Making Framework was developed and used for the aforementioned Palau Workshop. Guided by a series of questions, this is a voluntary framework, directing a step by step reflection for different constituencies on World Heritage and the tourism issue. Information is divided into three sections; a section on World Heritage issues, a section on planning and management, and a section on the national tourism issues affecting the World Heritage site. Questions build off the information in each chapter. For example, a key question, is, 'Why is our place unique, special and globally important?' and then "What kind of tourism do we wish to provide visitors to maintain this sense of place?" A follow-up question is "What are the complementary factors, awareness raising and pride, better planning, additional funding, etc., that can contribute to the maintenance and sustainability of that vision and what are the actions needed for their implementation?" For non-World Heritage areas, perhaps the workshop model could be modified and adapted providing an additional process for reflection.

4.7 Community Engagement

The aforementioned processes described in this chapter, suggest in practice, the need for robust community engagement to make motivations and interests explicit, increasing the potential to maximise the range of tourism benefits. There is an overarching logic here because besides the untapped benefits that tourism could help to facilitate there is the underlying and nagging knowledge that "while benefits can be derived by communities from tourism initiatives in which they have no involvement, it is not clear whether the appropriate, correct or most effective benefits are derived by the relevant community" (Simpson 2008).

The aforementioned processes and their implementation could also benefit from a shift in the style of community engagement. "Reviews of community development, beyond tourism, challenge the role of external consultants arguing for a shift away from consultant experts to facilitators and coordinators with longer term relationships with the communities in question" (Moscardo and Murphy 2014).

The community engagement skills that could support the implementation of the processes mentioned in this chapter are many and varied. These skills are described in detail in other materials and textbooks, and include the need for facilitation and listening skills as well as the ability to formulate and ask appropriate questions helping to identify purpose, needs, and to better target benefits. Several suggestions to begin to investigate this vast field of knowledge include:

- Community Places, 2 Downshire Place, Belfast, BT2 7JQ, Ireland – http://www.communityplanningtoolkit.org
- Department of Environment and Primary Industries, State Government, Victoria, Australia – http://www.dse.vic.gov.au/effective-engagement/toolkit
- The South African Tourism Planning Toolkit for Local Government January 2009 – http://www.kznded.gov.za

But community engagement doesn't necessarily mean successful outcomes. Local representatives may be appointed by government, and not elected by the people. Representatives may be members of the political parties forming the Government and not truly represent the interests of the local electorates. Public participation may be limited to very few people, the articulate and concerned individuals and leaders of organizations and community groups (Som et al. 2007). Contingencies for these and the multitude of possible complexities are needed and the work adapted as the processes move forward; all this the domain of agile learning organizations helping to make community engagement for expanded tourism integration a reality.

4.8 Conclusion

Someone once said that, "ideas should feel like affinities and not impositions" and the idea for this chapter has been to suggest approaches that students can hopefully relate to. The specific intent is to, in a positive fashion help move efforts ahead for widely integrating tourism into development.

However, the other intent is to provide information and ideas that are concretely optimistic to young people, because sustainable tourism, indeed the concept of sustainability, needs a shot in the arm, would benefit from a healthy measure of inspiration. But how can sustainable tourism be more inspirational? Maybe the example of sustainable architecture can serve as a metaphor for what is needed with the sustainable tourism movement.

If one asks oneself, what is sustainable architecture, one could say it is saving on energy costs, using environmentally friendly materials, and not harming the land where construction is taking place. These elements are certainly necessary. More inspiring however, might be adding, that the building in question reinforces your relationship with the place, encourages the education of your children, makes you healthier, makes you proud, and encourages inviting people in to become a greater part of the community. If you have all these elements combined in a building that is both beautiful and elegant you may have the makings of a more satisfied and delightful life. Perhaps this is the vision we need to keep in mind to inspire us to stay with it on the sustainable tourism issue.

References

Coffey International Development. 2013. http://us.coffey.com/international-development/news/loboc-tourism-and-community-benefit-from-australia-awards2013. Accessed 5 Mar 2014.

European Commission. 2013. Eurostat. http://epp.eurostat.ec.europa.eu/portal/page/portal/sdi/indicators. Accessed 12 Apr 2014.

European Commission. 2014. Erasmus Mundus. *Her.cul.es – Strengthening the attractiveness of European higher education in Heritage and Cultural Tourism*. http://www.herculesproject.eu/index.php/en/. Accessed 23 Apr 2014.

Himalayan Trust. 2014. http://www.siredmundhillary.com/projects.htm. Accessed 18 Mar 2014.

International Council on Monuments and Sites (ICOMOS). 1999. http://icomos.org/en/charters-and-texts. Accessed 10 Mar 2014.

McCool, S.F. 1999. *Making tourism sustainable: Sustainable tourism and what should tourism sustain: Different questions, different indicators*. Paper presented at international symposium on coastal and marine tourism, Vancouver, Apr 1999.

Mekong Tourism Coordinating Office. 2009. http://www.mekongtourism.org/site-t3/partners/development/ifc-mpdf/. Accessed 10 Mar 2014.

Moscardo, G., and L. Murphy. 2014. There is no such thing as sustainable tourism: Re-conceptualizing tourism as a tool for sustainability. *Sustainability* 6: 2538–2561.

Myrtle Beach Area Chamber of Commerce, Tourism Growth Initiative. 2011. http://www.tourismworksforus.com/tgi_about.html. Accessed 5 Mar 2014.

Pedersen, A., and H. Ceballos-Lascurain. 1990. *Nature-oriented tourism in the State of Guerrero, Mexico: Issues and recommended policies for local economic development*. Commission for the Study of International Migration and Cooperative Economic Development. Working paper, No. 59.

Pugh, C., and E. Wood. 2004. The strategic use of events within local government: A study of London borough councils. *Event Management* 9: 61–67.

Rebanks, J. 2008. *The economic gain: Research and analysis of the socio economic impact potential of UNESCO World Heritage Site Status*. Rebanks Consulting Ltd and Trends Business Research Ltd.

Rebanks, J. 2010. *What has World Heritage Site status ever done for us? The socio-economic impact potential of UNESCO World Heritage Site status*. Paper received from James Rebanks, Director, Rebanks Consulting Ltd., Cumbria.

Saarinen, J. 2013. Critical sustainability: Setting the limits to growth and responsibility in tourism. *Sustainability* 6: 1–17.

Simpson, M.C. 2008. Community benefit tourism initiatives – A conceptual oxymoron? *Tourism Management* 29(1): 1–18.

Som, P., M. Badaruddin, J. Jusoh, A. Marzuki, and M. Bahauddin. 2007. Community approaches in tourism planning at grass root level. *TEAM Journal of Hospitality and Tourism* 4(1): 56–68.

The Global Sustainable Tourism Council (GSTC). 2013. http://www.gstcouncil.org/sustainable-tourism-gstc-criteria/criteria-for-destinations.htm. Accessed 21 Mar 2014.

Tourism Queensland. 2014. How tourism benefits communities. http://www.tq.com.au/resource-centre/index.cfm. Accessed 10 Mar 2014.

Waddilove, A., and H. Goodwin. 2010. *Tourism in transition? Incorporating tourism into the transition mode*. International Centre for Responsible Tourism, Occasional Paper No. 18, Leeds Metropolitan University.

World Monuments Fund. 2014. http://www.wmf.org/project/suchitoto-city. Accessed 12 Mar 2014.

Chapter 5
Strategic Community Participation in Sustainable Tourism

Susan Snyman

Abstract Tourism has become an increasingly complex phenomenon, with political, economic, social, cultural, educational, biophysical, ecological and aesthetic dimensions. In order for tourism to be sustainable it must bring direct, as well as indirect, benefits to host communities. The aim is to provide an important means and motivation for communities to care for and maintain their natural and cultural heritage and cultural practices. In this process, there are diverse challenges, as well as opportunities, facing communities developing or engaging in tourism. This chapter provides a framework to assist communities in developing tourism in their area effectively, efficiently, equitably and sustainably.

Keywords Community participation • Development • Tourism • Benefit sharing • Sustainability

5.1 Introduction

Tourism is frequently seen as a panacea and as an economic activity which will end poverty and solve socio economic problems. It is certainly not a panacea, but can go a long way to being a catalyst for socio-economic development and can serve to alleviate poverty and promote development in a number of destinations. There are, however, some caveats which attach to this and which are important to bear in mind, particularly in destinations new to tourism or starting to develop tourism. Communities are not homogenous and not all residents in a destination will support the development and integration of tourism into their community. It is important, therefore, to assess the opinions and attitudes of residents to tourism development to ensure that any development is managed in such a way as to maximise support and buy-in from the majority of the community. Part of this process should include an assessment of the technical and financial capabilities in the community, as well

S. Snyman (✉)
Environmental Policy Research Unit, University of Cape Town and Wilderness Safaris, South Africa
e-mail: suesnyman@gmail.com

S.F. McCool, K. Bosak (eds.), *Reframing Sustainable Tourism*,
Environmental Challenges and Solutions 2, DOI 10.1007/978-94-017-7209-9_5

as a thorough assessment of what exactly the community has to offer tourists and how best to market this and ensure that it is of high quality and is competitive. The particular form of tourism chosen for a destination has impacts not only on the host destination, but there are also potential broader developmental outcomes which can benefit the destination (Telfer and Sharpley 2008). Gunn and Var (2002 in Telfer and Sharpley 2008: 81) identify the goals for better tourism development as enhanced visitor attractions, improved economy and business success, sustainable resource use and community and destination integration. This chapter will elaborate on each of these goals, but the discussion starts with a brief overview of some of the benefits and risks associated with tourism and which need to be considered before a community engages in tourism.

5.2 Benefits for Communities Associated with Tourism Development

Some benefits associated with tourism include that it is inextricably linked to the environment, in terms of the attractions it offers to tourists, the quality of experience that can be provided, and the industry's impact from the use of resources to sustain it (Spenceley 2010: 20). Tourism can provide opportunities for tourists to learn more about the societies and values of the people they interact with and, through this, can ensure the preservation of local, cultural and historical traditions. Tourism revenues can be used to preserve historical sites and the tourism activity itself can revitalise local interest in history and traditional practices such as cuisine and entertainment (Spenceley 2010: 22). An innovative destination which has spearheaded the importance of society and culture in tourism is the Kingdom of Bhutan, on the edge of the Himalayas (Spenceley 2010). Another example is the Damara Living Museum in Namibia, which preserves the local Damara culture and offers tourists the opportunity to learn more about their cultural traditions. A possible reason for the success of both of these destinations is that population densities are relatively low and there has been focus on a common goal within the communities, with everyone working towards the same goal, i.e. a unified, structured approach to tourism.

It is important that a community decides together on what aspects of their destination they would like to incorporate into the tourism product, e.g. cultural history, natural resources, etc. Once this goal is established there can be a unified approach to achieving it.

The benefits of ecotourism are not, however, unambiguously positive, though some authors (Lapeyre 2011) have argued that it can be a 'solution' in rural areas as it helps achieve the triple bottom line of sustainability:

(i) it creates full-time, as well as casual, seasonal and contract employment and therefore income opportunities for poor households in remote areas (economic sustainability);

(ii) it encourages individual and, in some cases, collective local empowerment through skills training and development and the sharing of decision-making (social sustainability);
(iii) it generates income for both the community as a whole, and for individuals, that can be used to support conservation costs as well to encourage positive conservation behaviours (environmental sustainability).

It was observed by Snyman (2013) and earlier noted by Adams and Infield (2003) that income generated from tourism can however lead to a dynamic of competition of its own, as different stakeholders attempt to dominate access to the available revenue streams. The private sector operator wants to maximise profits, as would the community and any related government departments; 'best practice' involves maximising the benefits of all involved, or at least satisficing.

5.3 Challenges and Risks for Communities Associated with Tourism Development

Tourism is often put forward as a tool for conservation and sustainable development (Ceballos-Lascuráin 1998, as cited in Tsaur et al. 2006; Koelble 2011). Despite reducing risk for local communities through incomes earned, tourism has its own risks, some potentially more problematic than other land uses; for example, sensitivity to exchange rates and the oil price, natural disasters, politics and health scares, all of which can destroy tourism in a destination.

Prospect theory and loss aversion (Kahneman and Tversky 1979) suggest that the negative aspects of tourism may be more obvious to local residents and they may wish therefore to avoid these more than to receive benefits, since some of the benefits derived from tourism appear as free goods and may therefore go unnoticed (Snyman 2013). The costs however, which appear intermittently, are likely to be more direct and therefore obvious to households. This can threaten community acceptance of tourism (Snyman 2013). As an example, incidents of crime or overcrowding due to tourism and their direct impacts on households' livelihoods and welfare are more likely to be strongly remembered by households, irrespective of whether or not they are concurrently receiving benefits from tourism, i.e. improved roads and communication (Snyman 2013).

Naturally, the social, political, environmental and economic context of a destination will influence the impacts any tourism development will have. Telfer and Sharpley (2008) stress that what is positive in one destination, may not be in another (for example, an increase in tourist numbers in a city may have a positive impact through increased revenues with no negative impacts, but a similar increase in tourist numbers to an ecologically-sensitive site in the Okavango Delta in Botswana may increase revenues, but have a negative impact environmentally).

A common obstacle to broadening linkages around tourism operations is the inadequacy of skills, leaving communities unable to provide the required goods

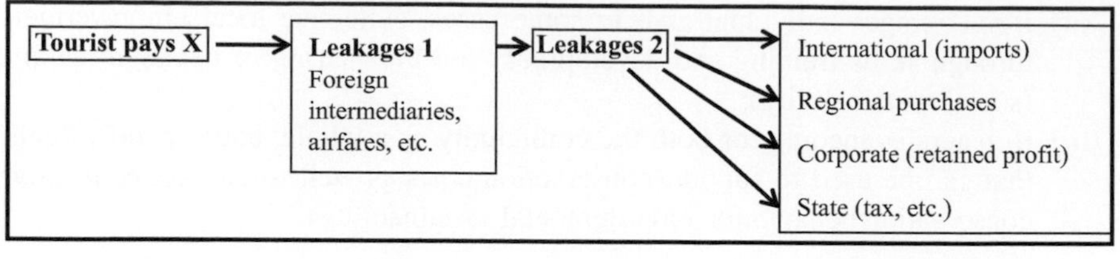

Fig. 5.1 Leakages from tourism (Snyman 2013)

and services, unaware about tourist demands and therefore the types of goods and services to provide (Snyman 2013). This problem has been widely discussed (see for example Ogutu 2002; Epler Wood International 2004; Rogerson 2006) and Mitchell and Ashley (2010) stress the need for a multi-faceted approach to promote and increase linkages.

There has been much debate regarding the 'leakage' from tourism operations in rural areas. Figure 5.1 shows possible leakages from tourism. Meyer (2008: 561) defines leakage as the percentage of the price of a holiday paid by tourists that either leaves the destination in payment for imports or as expatriated profits, or that never reaches the destination due to the involvement of foreign intermediaries. Leakages can be identified and measured by assessing the supply of goods and services that are being imported to fill market needs and, from there, looking for local alternatives (Snyman 2013).

One area of difficulty for tourism development is, therefore, to balance all the costs and benefits of such development. Frechtling (1994) emphasised the need to consider opportunity costs, i.e. the returns on the highest-value alternative resource use should be subtracted from the net benefits of tourism to obtain its true economic benefit to the economy (Snyman 2013).

Careful consideration needs to be given to the possibility of a decline in tourism over time, or for a particular reason (for example prior to an election, a natural disaster, etc.), etc. Contingency plans should be in place. Surviving declines in tourism can be accomplished through the diversification of tourism products, encouraging investment, high quality vocational training, and aggressive marketing and promotion (Spenceley 2010: 6).

Based on the above discussion of the various risks and opportunities which communities looking at developing tourism face, the next section suggests a framework for managing tourism development in a destination using Fig. 5.2 below as a guideline for a sustainable tourism development process.

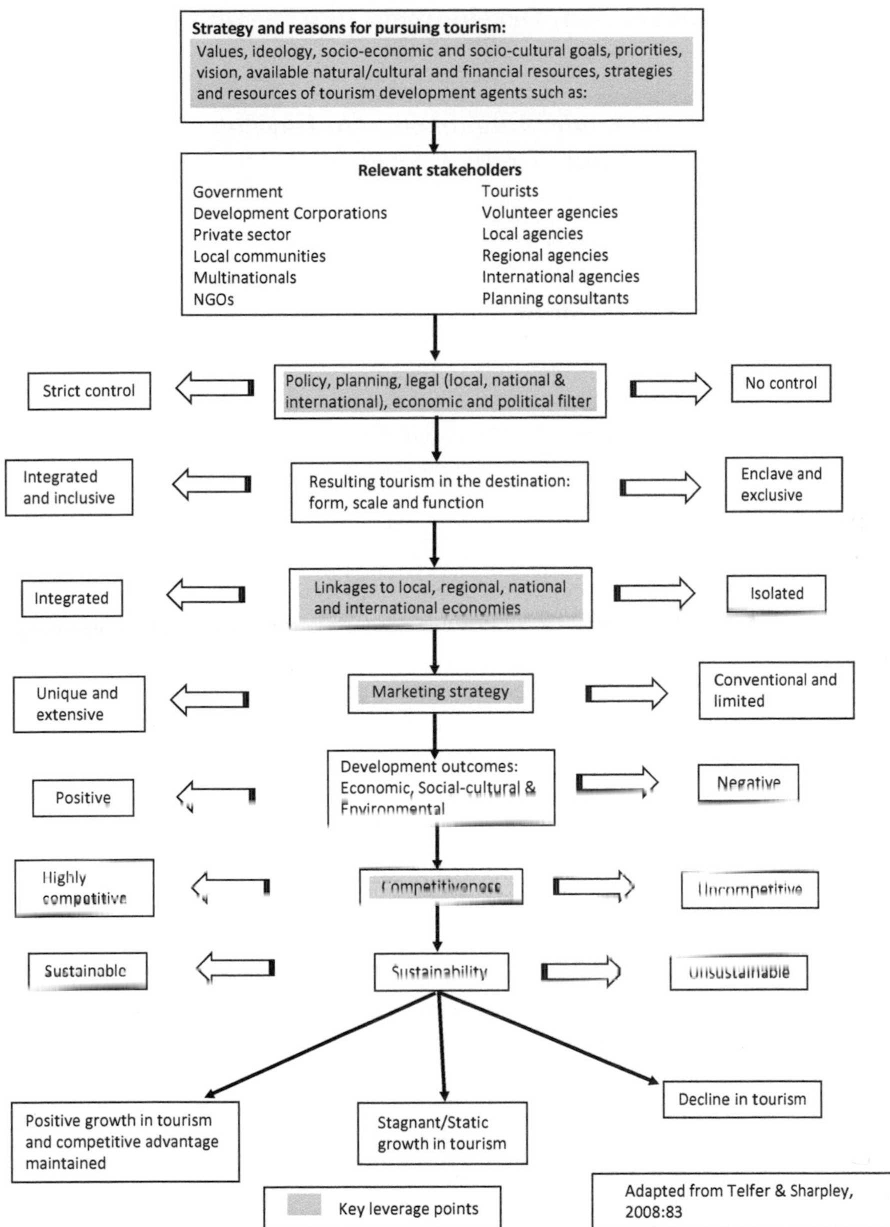

Fig. 5.2 Sustainable tourism development process

5.4 Managing Tourism Development in a Destination: A Suggested Framework

5.4.1 Create Stakeholder Engagement to Develop Ownership in the Community Tourism Strategy

An important issue to consider is who the major stakeholders are, how are they defined, and what the roles and responsibilities of each will be. Clarity with regards to roles is important in reducing conflict in the long run and in ensuring that tourism in the destination is more sustainable. The major stakeholders include, amongst others, government, residents (community), private sector and the tourists themselves. The collaboration of major stakeholders in destinations can be effective in providing synergistic financial, technical and local support (Spenceley 2010). On the other hand, a lack of collaboration can lead to conflict and discord which can undermine the success of tourism development in a destination.

According to Shani and Pizam (2012) there is a unanimous agreement that the most critical stakeholders that should be considered in the tourism planning literature are the local residents (deAraujo and Bramwell 1999; Sautter and Leisen 1999 in Shani and Pizam 2012). The process of taking the host community's perspective into account can be seen as a means of balancing the needs of the local residents with those of other prominent stakeholders such as tourism developers, businesses, environmental groups, and municipal and governmental authorities (Tosun 2000; Williams and Lawson 2001 in Shani and Pizam 2012: 548).

The issue of choice is central to the tourism debate. It is important therefore to differentiate between constraints on land use imposed on communities, and those entered into voluntarily. An observation of Snyman (2013 and earlier made by Eagles et al. 2002) is that negative impacts of tourism tend to be more common when communities are not given choices, i.e. when they have no control over their involvement in the tourism activities; this can be mitigated when local communities are involved in the planning and management of tourism in their destination (see Snyman 2012, 2013). This includes including them in discussions and negotiations and the actual establishment of tourism in the destination.

Governments need to create a stable political and economic climate, secure land tenure, safety for visitors, favourable conditions for investors, and ensuring a good reputation and 'brand' for their tourism destinations (Spenceley 2010). These conditions are created through government led policies, plans and laws that guide and regulate the sector (Spenceley 2010). Noteworthy initiatives in Africa include Namibia's conservancy programme, which gives rural communities the rights to use wildlife on their lands; Zanzibar's Strategy for Growth and Poverty Reduction, and the Seychelles Ecotourism Strategy (see Spenceley 2010 for case studies on each of these examples). Other innovative approaches in Sub-Saharan Africa include the

South African National Parks programme for public-private partnerships (PPPs) in national parks, responsible tourism policies developed in South Africa and the Gambia, and Botswana's policy to promote high-value, low impact tourism (see case study 2 in this chapter) (Spenceley 2010).

Another major stakeholder is the community, whose role is largely in the provision of goods and services, the management of tourism development and the provision of access to resources (natural, historical, cultural, etc.) in their destination.

The private sector has an important role to play in tourism development in that they can contribute skills, expertise and capital to communities in the early stages of tourism development. The private sector has impacts on the local economy, through employment and procurement; on the natural environment from the way in which they develop infrastructure, conduct their tours and use resources; and on society and culture in which they operate and sometimes commercialise (Spenceley 2010: 15).

High-end private sector tourism enterprises can generate significant returns in destinations including local employment and training, capital investment in infrastructure, local procurement, conservation and corporate social responsibility (Spenceley 2010). However, to attract such investors to a destination requires an enabling environment including stable land tenure, political stability, access, suitable infrastructure and support, suitable attractions in the destination, natural or other, and medium- to long-term commitment (Spenceley 2010). This is elaborated on in the next section.

Tourists are stakeholders whose particular needs and wants play an important role in determining a community's tourism strategy and what they will or will not offer as part of their tourism product. A thorough understanding of these needs and wants is an important first step for a community pursuing tourism development in their destination. This can be achieved through a questionnaire survey, face-to-face interviews with tourists, or through a detailed desktop survey of successful tourism products and research into the reasons for their success.

An important part of stakeholder participation which has been argued by numerous authors is that any successful tourism project requires regular communication and dialogue with involved communities (Armstrong 2012; Nepal and Weber 1995, as cited in Bruyere et al. 2009; Chandralal 2010). The view here is that meaningful dialogue and the sharing of information between stakeholders helps to develop mutual trust and to build social capacity in communities (Snyman 2013). The form and frequency with which it takes place would be determined by the relevant cultural and social norms in the area of operation and the need to align stakeholder objectives, especially when expectations about tourism's results differ. In such cases, communication would not only be needed in the development phases, but should be ongoing, with regular information-sharing, to minimise distrust and lack of interest amongst communities (Snyman 2013).

5.4.2 Ensure an Enabling, Supportive Regulatory Environment for Tourism Investment, Development and Growth

The importance of sound policy and regulatory frameworks that create stability and clarity related to tourism development cannot be overstated (Spenceley 2010), particularly for private sector operators looking to invest in a destination.

Government has an important role to play in terms of legislation governing tourism (Tourism Acts, etc.) and ensuring that there is an overall enabling framework within which tourism will develop in a destination and which will encourage private sector investment. This framework will determine what restrictions are placed on development, what environmental impact assessments are required, whether or not visas/taxes, etc. will affect tourism numbers and how the private sector will be involved. It is important that local communities are aware of all government regulations surrounding the tourism industry in their destination. This includes all health and safety, as well as environmental regulations which need to be adhered to and are particularly pertinent in a number of tourism activities related to nature and the outdoors.

There are at times extenuating circumstances which tourism operators have no control over. These circumstances may have positive or negative impacts on tourism in an area. Case Study 1 illustrates an example of potential negative impacts. Such a situation will determine the level of private sector investment in tourism in an area/country, which will obviously impact on communities in the area as well.

Once the regulatory environment is suitable for tourism development, the community needs to make decisions related to what form or forms of tourism are best suited to their destination in the long term in order to meet socio-economic and environmental goals (Telfer and Sharpley 2008). Each type of tourism development will bring different tourists, with varying levels of disposable income and expectations, along with different opportunities for locals to participate in the tourism economy (Telfer and Sharpley 2008: 81). There is a clear need to choose a tourism development strategy which is appropriate to attaining desired community goals which, therefore, have to be clearly articulated and understood by all relevant stakeholders.

Related to governance and establishing a suitable regulatory environment is the establishment of a Tourism Board/Association. This is an important starting point for a community and will provide structure to the development of tourism in the destination, and allow for the development of a framework focusing specifically on the development of tourism in a particular destination. Tosun (2006: 498) found in their study in Turkey that the majority of respondents support the idea that local people should be consulted about local tourism development issues and the second most popular option was the idea that a committee elected by the public especially for developing, managing and controlling tourism development should decide on all aspects of local tourism development.

The Tourism Board should include community members, as well as private sector investors. It should have a 2- to 5-year Strategic Plan for the development of tourism in the destination. Understanding tourism, what will be offered, how much, when the peak seasons will be, what activities will be offered, by who, when, and for how much are all important issues to discuss and for all stakeholders to agree on.

There should be specific, clearly stated requirements to be members of the Tourism Association. Membership fees should not be excessive, to avoid excluding certain residents, as this may cause conflict and reduce the socio-economic impacts of tourism in the destination. It must be clear to members what they will receive in return for their membership fee, as this will encourage residents to be members and to align with the town's strategic plan for tourism development in the destination. Prior to tourism developing in a destination, it is also important to have set standards of what will be accepted in terms of accommodation facilities, activities offered, etc., to ensure the maintenance of good quality products and services.

Conflict management systems should be in place prior to tourism developing to ensure that should conflict between stakeholders arise at a later stage, there are mechanisms in place to manage it. In the case of a conflict of interests between different stakeholders, the optimal balance between them should be sought during the decision-making process (Hasnas 1998 and Ogden and Watson 1999 in Shani and Pizam 2012).

In summary, solid, accountable and transparent governance structures with clearly defined roles are, therefore, important for the long-term success of tourism in an area (Snyman 2013). Support institutions that emerge to coordinate and regulate the interactions between stakeholders can also assist with education, training, skills development and, in some cases, conflict resolution.

5.4.3 *Ensure Needed Capacities and Skills for Servicing Tourists Exist*

Observed by Snyman (2013) and highlighted by numerous other authors (Armstrong 2012; Boudreaux and Nelson 2011; Coria and Calfucura 2012) is the problem of communities that need training in administration, financial reporting and strategic planning as well as other skills such as bookkeeping, negotiating contracts and improving communication and marketing capabilities.

An important part of ensuring that tourism development in a community is competitive is, therefore, to ascertain what skills, goods and services community members have to offer to tourism, and to focus growth and development on incorporating already existing goods and services. Ashley (2000) emphasised that matching the design of tourism operations to local livelihoods requires a thorough understanding of people's livelihood strategies, needs and alternatives. This is important in terms of understanding how tourism can and should develop to maximise positive impacts and local linkages.

The private sector has an important role to play in terms of capacity building and skills training and development as they frequently have internal resources and expertise which can be used to upskill local community members. Snyman (2013) found that out of 385 tourism employees interviewed in six southern African countries for 63 % of them their current job in tourism was their first permanent job, highlighting the importance of skills acquired in these jobs. One objective of private sector investment in education and training is to differentiate one ecotourism operation from others in the area as service quality is often an outstanding feature in the high-end tourism market (Snyman 2013). Training and uplifting staff are also intended to increase job satisfaction and therefore staff retention, thereby lowering overall training costs and improving service levels (Snyman 2013).

5.4.4 Develop and Sustain a Competitive Advantage

Tourism is becoming an increasingly competitive industry and a community, therefore, has to ensure that they develop a competitive advantage. This can be done in a number of different ways, including through ensuring they have a unique product, a high quality product or service at a reasonable price, or appealing to a specific target market, e.g. birdwatchers. They also have to decide how they will maintain their competitive advantage over time. The earlier a community gives these factors consideration, the more likely it will be that tourism will be sustainable and successful. Case study 2 illustrates the high cost/low volume tourism strategy successfully chosen by Botswana. This strategy has resulted in significant returns from tourism for Botswana as well as the maintenance of environmental integrity.

As part of this, Local Direct Marketing Organisations (DMOs) should have a clearly defined Marketing Plan which should include details of how the destination will be marketed; what are the Unique Selling Points (USPs), i.e. what differentiates this destination from others; what will encourage tourists to visit; how do you want to be perceived (for example, as a backpackers destination or as a high-end tourism destination); what is the competitive advantage that will ensure tourists visit the destination, and how can you maintain a competitive advantage over time.

Local DMOs in a destination should consider a number of important questions prior to developing tourism in a destination; these include:

- What are the desired social, economic and environmental outcomes of the tourism development in your destination?
- What tourism, environmental and other relevant regulations, laws, etc. exist in your destination?
- How will these rules and regulations impact on tourism development in the destination?
- What mechanisms are in place to encourage private sector investment in the destination?
- How will membership of the Tourism Association be defined?
- What standards of and requirements for membership are going to be set?
- How will these standards be maintained, monitored and enforced?
- What are the unique selling points of the destination and what are the best ways to market them?
- How will those who do not uphold tourism standards set (these should also include environmental standards) be managed?

- What are the long-term goals of tourism development (i.e. job creation, increased incomes, preservation of culture, etc.) in the destination?
- What competitive advantage is the destination trying to achieve and how will this be maintained over time?

The DMO should have an up-to-date strategic plan, with community buy-in (need to ensure that all stakeholders are on the same page), which is based on research related to whether or not the community is ready to pursue tourism development and what form of tourism will best achieve community development goals, whether these be to increase high school graduation rates, improve social standards in the destination or to maximise labour income. The strategic plan should include innovative, creative ways to market unique aspects of the destination in order to maintain a competitive advantage.

5.5 Necessary Steps to Ensure That a Destination Maximises on Their Returns from Tourism Development

There are a number of necessary steps to ensure that a destination maximises on their returns from tourism development. These include:

1. Research into the strengths and weaknesses of the destination in terms of developing a competitive advantage;
2. Integration of these results into the destination's tourism strategy;
3. Research into the available goods, services and skills available in the community;
4. Integration of these results into the design of the tourism product;
5. Use of tourism consultants in the research and initial tourism development phases;
6. Skills trainings and workshops to increase human capital and empower local communities to manage and operate tourism in their destination;
7. Regular appraisals of tourism development, particularly related to the destination's specific goals for tourism development, and;
8. Assessment of whether or not the outcomes of tourism development are in line with those originally proposed;
9. Adjustments to the tourism development strategy based on the appraisals and assessments.

Case study examples from Spenceley (2010: 19) in Africa illustrate the impacts of different tourism strategies:

- **Number of tourists and their expenditure**: The number of tourists visiting the Seychelles and business travellers in Nairobi is similar, but the average expenditure is much higher in the Seychelles (US$230 vs. $128 per day).
- **Length of stay and spend per day**: Encouraging tourists to stay longer, and spend more money locally each day, generates more income in destinations. For example, an average expenditure US$2303 per trip (or $230 per day) in the Seychelles for a 10 day trip, generates three times the revenue of a seven day business trip to Nairobi, at $762 per trip (or $108 per day).

- **Investment value**: Collectively two private sector companies (&Beyond and Wilderness Safaris) have invested almost as much money in their tourism enterprises, as has been invested in all 59 conservancies in Namibia (US$16.5 million vs. $19 million). This is an impressive achievement from two private operators, and provides support for the importance of a positive business enabling environment where the private sector can excel.
- **Number of jobs**: Although Mount Kilimanjaro has roughly half the number of tourists as Pays Dogon in Mali, and almost 70 % of the average trip spend, this destination generates 52 times the number of jobs. However, these are seasonal, informal jobs, and salary levels are low.

It is therefore important that communities are clear what their tourism strategy is and how they plan to achieve it.

5.6 Conclusion

The choice of tourism strategy to implement in a destination has important implications for the results that will be achieved, i.e. focusing on volume, expenditure, duration of stay, capital investment, etc. Destinations choosing to pursue tourism should be clear on their strategy and the reasons for pursuing tourism should be clear and based on community level goals discussed in this chapter.

In summary, Spenceley (2010: 6) suggests a series of recommendations that may assist destinations in overcoming constraints to sustainable tourism development; these include (adapted from Spenceley 2010: 6):

- Development of enabling policies, that are based on sound research and participatory development processes;
- Development of appropriate instruments and programs to implement and regulate those policies consistently (e.g. focussing on yield, rather than numbers of tourists);
- Simplifying and supporting the development and operation of business through suitable licensing and regulatory instruments;
- Vigorously tackling local and national corruption and poor governance by providing transparent and equitable solutions;
- Providing trust, space, time and an enabling business environment for innovation by the private sector;
- Monitoring and evaluation of the impact of tourism, as well as other relevant, policies, and providing mechanisms for feedback and adaptation;
- Targeted capital infrastructure development to support tourism, related to demand, and simultaneously enhancing the destination for residents through improved facilities for healthcare, education, transport and sanitation;
- Creating incentive and taxation instruments that support, rather than punish commercial success;

- Recognising that the people living in tourism destinations are an integral part of the asset;
- Adopting participatory processes for planning and decision making with local people;
- Providing mechanisms to ensure living, or minimum wages, across the tourism sector, in partnership with the private sector;
- Promoting value for money in tourism products and destinations, coupled with high quality service and experiences, i.e. creating a competitive advantage;
- Establishing strong market linkages between the destination and source markets and encouraging investment in marketing and promotion;
- Promoting strong local value chains, so that local businesses can overcome barriers to engaging in tourism markets, and sell their goods and services to the tourism sector;
- Monitoring and evaluating the economic and financial returns to society and local people;
- Ensuring adequate planning, design and location for tourism development, which is cognisant of the impacts on the local environment and resource use;
- Avoiding negative environmental impacts where possible, and mitigating damage when it occurs;
- Providing access to information and technical assistance to support tourism development;
- Providing access to vocational training for local people in hospitality and tourism (including guiding, marketing and craft development);
- Protecting the rights of workers to safe and healthy working conditions;
- Using tourism to conserve, rehabilitate and re-invigorate cultural and natural heritage and traditions;
- Tackling and resolving conflicts as they arise, and trying to find win-win solutions.

Tourism development does not just occur, it needs to be carefully constructed and aligned to the goals and preferred outcomes of the destination. Today more than ever, tourism is a highly competitive industry with thousands of communities, destinations and tourism businesses competing with one another to attract tourists to their destination. Development of tourism should, therefore, ensure that the destination is competitive, with a high quality, unique product, offered with an understanding of the tourism market, at a reasonable price and in so doing, maximising positive benefits for the tourist and the community alike.

Case Study 1: Africa: Limiting Tourism Development
Certain countries, particularly in Africa, do not provide the ideal enabling environment to encourage investment in the tourism sector, particularly private sector investment. Restrictive tax and visa policies discourage private sector investment and make it difficult for tourists to gain access. Flight schedules, road and communication networks, a lack of good quality accommodation and tourism activities all hinder tourism development and growth. Other problems include governance issues, corruption, political instability which may result in safety issues, distances

from tourist source markets, access to quality goods and services for tourism and high competition in the ecotourism sector, which is the main form of tourism in many destinations in Africa.

Case Study 2: Botswana's High Cost/Low Volume Policy

Botswana's tourism strategy has always been to expand low volume/high cost ecotourism (Lepper and Goebel 2010; Sammy and Opio 2005). This policy is designed to limit the negative impacts of mass tourism as well as ensure an exclusive experience. It therefore keeps costs and ecological impacts low and maintains a high willingness to pay (WTP) among the small number of users (Snyman 2013). For this, Government has set aside more than 17 % of all available land for National Parks and wildlife reserves and a further 22 % as wildlife management areas (WMAs) (Leechor and Fabricius n.d.).

Although tourism makes up a relatively small portion of Botswana's GDP (3.4 % in 2005/2006) it is likely to prove extremely important to the country's future growth and, with the declining value of diamonds, the government is looking increasingly towards tourism (Tourism Statistics Botswana 2007). According to the Botswana Department of Tourism, tourism has grown at an average of 8.4 % per annum since 1994. Between 2000 and 2009 the number of tourists arriving in Botswana grew by over 50 % (Botswana Tourism Research 2011). According to the WTTC (2012), the direct contribution of travel and tourism to GDP in Botswana was expected to be BWP3.03 million (2.4 % of total GDP) and it was expected to directly support 18,500 jobs (3.1 % of total employment) in 2011 (Fig. 5.3).

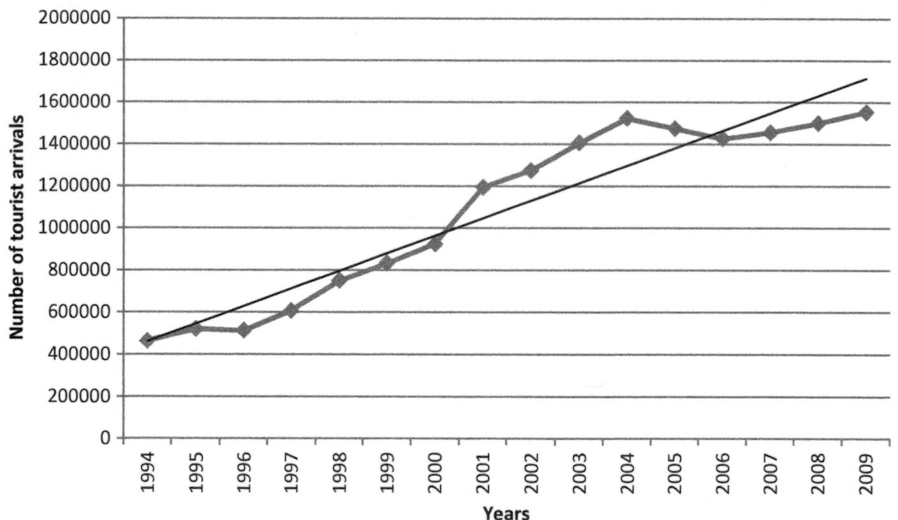

Fig. 5.3 Botswana tourism statistics, 1994–2009 (Source: Botswana Department of Tourism 2009)

References

Adams, W.M., and M. Infield. 2003. Who is on the gorilla's payroll? Claims on tourist revenue from a Ugandan National Park. *World Development* 31(1): 177–190. doi:10.1016/S0305-750X(02)00149-3.

Armstrong, R. 2012. *An analysis of the conditions for success of community-based tourism enterprises*. International Centre for Responsible Tourism, occasional paper OP21. http://www.artyforum.info/RTD/OP21RebeccaArmstrong.pdf. Retrieved 2 Feb 2012.

Ashley, C. 2000. *The impacts of tourism on rural livelihoods: Namibia's experience*. Overseas Development Institute, working paper 128. http://www.odi.org.uk/sites/odi.org.uk/files/odi-assets/publications-opinion-files/2754.pdf. Retrieved 18 Dec 2008.

Botswana Tourism Research. 2011. http://botswanatourismresearch.org/. Accessed 17 May 2012.

Botswana Tourism Statistics. 2007. Tourism statistics: 2006–2009.http://www.mewt.gov.bw/uploads/files/TOURISM%20STATISTICS%202006-2009_1.pdf. Retrieved 11 May 2014.

Boudreaux, K., and F. Nelson. 2011. Community conservation in Namibia: Empowering the poor with property rights. *Economic Affairs* 31(2): 17–24. doi:10.1111/j.1468-0270.2011.02096.x.

Bruyere, B.L., A.W. Beh, and G. Lelengula. 2009. Differences in perceptions of communications, tourism benefits, and management issues in a protected area of rural Kenya. *Environmental Management* 43: 49–59. doi:10.1007/s00267-008-9190-7.

Chandralal, K.P.L. 2010. Impacts of tourism and community attitudes towards tourism: A case study in Sri Lanka. *South Asian Journal of Tourism and Heritage* 3(2): 41–49. Retrieved from http://www.academia.edu/822311/Impacts_of_Tourism_and_Community_Attitude_towards_Tourism_A_Case_Study_in_Sri_Lanka.

Coria, J., and E. Calfucura. 2012. Ecotourism and the development of indigenous communities: The good, the bad and the ugly. *Ecological Economics* 73: 47–55. doi:10.1016/j.ecolecon.2011.10.024.

Eagles, P.F.J., S.F. McCool, and C.D.A. Haynes. 2002. *Sustainable tourism in protected areas: Guidelines for planning and management*. Gland/Cambridge, UK: IUCN, xv + 183 pp. http://ecosynapsis.net/RANPAold/Contenido/MainPages/preAmac/articulosPDF/sustainable_tourism_in_pa_guidelines.pdf. Retrieved 23 Mar 2011.

Epler Wood International. 2004. *Evaluating ecotourism as a community and economic development strategy*. The EplerWood report. www.eplerwood.com. Retrieved 10 Feb 2012.

Frechtling, D.C. 1994. Chapter 33: Assessing the impacts of travel and tourism – Measuring economic costs. In *Travel, tourism and hospitality research: A handbook for managers and researchers*, 2nd ed, ed. J.R. Brent Ritchie and C.R. Goeldner. New York: Wiley.

Kahneman, D., and A. Tversky. 1979. Prospect theory: An analysis of decision under risk. *Econometrica* 47(2): 263–292. http://www.jstor.org/stable/1914185

Koelble, T.A. 2011. Ecology, economy and empowerment: Ecotourism and the game lodge industry in South Africa. *Business and Politics* 13(1), 3. doi:10.2202/1469-3569.1333.

Lapeyre, R. 2011. The Grootberg lodge partnership in Namibia: Towards poverty alleviation and empowerment for long-term sustainability? *Current Issues in Tourism* 14(3): 221–234. doi:10.1080/13683500.2011.555521.

Leechor, C., and M. Fabricius. n.d. *Developing tourism in Botswana: Progress and challenges*. Discussion draft commissioned as a background study to the World Bank-BIDPA Botswana Export Diversification Study. http://dspace.cigilibrary.org/jspui/bitstream/123456789/31841/1/TheTourismSector.pdf?1. Retrieved 3 May 2013.

Lepper, C.M., and J.S. Goebel. 2010. Community-based natural resource management, poverty alleviation and livelihood diversification: A case study from northern Botswana. *Development Southern Africa* 27(5): 725–739. doi:10.1080/0376835X.2010.522834.

Meyer, D. 2008. Pro-poor tourism: From leakages to linkages. A conceptual framework for creating linkages between the accommodation sector and 'poor' neighbouring communities. *Current Issues in Tourism* 10: 558–583. doi:10.2167/cit313.0.

Mitchell, J., and C. Ashley. 2010. *Tourism and poverty reduction: Pathways to prosperity*. London: Earthscan.

Ogutu, Z.A. 2002. The impact of ecotourism on livelihood and natural resource management in Eselenkei Amboseli Ecosystem, Kenya. *Land Degradation and Development* 13: 251–256. doi:10.1002/ldr.502.

Rogerson, C.M. 2006. Pro-poor local economic development in South Africa: The role of pro-poor tourism. *Local Environment* 11(1): 37–60. doi:10.1080/13549830500396149.

Sammy, J., and C. Opio. 2005. Problems and prospects for conservation and indigenous community development in rural Botswana. *Development Southern Africa* 22(1): 67–85. doi:10.1080/03768350500044644.

Shani, A., and A. Pizam. 2012. Community participation in tourism planning and development. In *Handbook of tourism and quality-of-life research: Enhancing the lives of tourists and residents of host communities*, International handbooks of quality-of-life, ed. M. Uysal, 547–564. Dordrecht/Heidelberg/London/New York: Springer. doi:10.1007/978-94-007-2288-0_32.

Snyman, S. 2012. Ecotourism joint ventures between the private sector and communities: An updated analysis of the Torra Conservancy and Damaraland Camp partnership, Namibia. *Tourism Management Perspectives* 4: 127–135. doi:10.1016/j.tmp.2012.07.004.

Snyman, S. 2013. *High-end ecotourism and rural communities in southern Africa: A socio-economic analysis*. Dissertation submitted for the fulfilment of the Doctor of Philosophy, School of Economics, University of Cape Town.

Spenceley, A. 2010. *Tourism product development interventions and best practices in sub-Saharan Africa. Part 1: Synthesis report*. Report to the World Bank, 27 Dec 2010.

Telfer, D.J., and R. Sharpley. 2008. *Tourism and development in the developing world*. Abingdon, Oxon: Routledge.

Tosun, C. 2006. Expected nature of community participation in tourism development. *Tourism Management* 27: 493–504. doi:10.1016/j.tourman.2004.12.004.

Tsaur, S.-H., Y.-C. Lin, and J.-H. Lin. 2006. Evaluating ecotourism sustainability from the integrated perspective of resource, community and tourism. *Tourism Management* 27: 640–653. doi:10.1016/j.tourman.2005.02.006.

World Travel and Tourism Council (WTTC). 2012. Travel and tourism: Economic impact 2012:Botswana. http://www.wttc.org/site_media/uploads/downloads/botswana2012.pdf. Retrieved 11 May 2014.

Chapter 6
Framework for Understanding Sustainability in the Context of Tourism Operators

Kelly Bricker and Rosemary Black

Abstract Tour operators are an important component of the tourism system. They link tourists and destinations, and as a result have influence on sustainability within destinations. Tour operators are also part of a larger system that include direct involvement in all aspects of a tourist experience, such as accommodations, attractions, food services, and transportation. The scale of tourism operators span from very localized to international and deliver a diverse array of services. However, increasingly tour operators face challenges in controlling aspects of the tourism product. Because tour operators influence many aspects of a destination and how visitors might perceive the destination, the can maximize aspects of sustainability including: social and economic benefits; benefits to cultural heritage; benefits to the environment, and enable partnerships. A framework for sustainable tourism operators captures the variability and potential role and influence tour operations have within the industry overall.

Keywords Sustainable tourism • GSTC • Criteria • Benefits • System • Resilient communities

6.1 Tour Operator Roles and Responsibilities

> There was a rustling about in the bush just in front of us . . . our group, only 8 people in total were under the leadership of two guides who appeared to be well versed in everything that surrounded us, from beetles to leopards, and ecosystems to surrounding villages. So when our guides instructed us to find the nearest tree, so they could "encourage" the mating rhinos to give up their efforts and run by us, in order to provide us with a closer view, my trust in their competence was seriously questioned. Nevertheless, as a point of survival, my friend and I clamored to the nearest low-hanging branch, as instructed and held on tight. What

K. Bricker (✉)
Department of Parks, Recreation, and Tourism, University of Utah, Salt Lake City, UT, USA
e-mail: Kelly.bricker@health.utah.edu

R. Black
School of Environmental Sciences, Charles Sturt University, PO Box 789, Albury, NSW 2640, Australia
e-mail: rblack@csu.edu.au

S.F. McCool, K. Bosak (eds.), *Reframing Sustainable Tourism*,
Environmental Challenges and Solutions 2, DOI 10.1007/978-94-017-7209-9_6

> we soon realized was the branch was smooth to touch on the underside, and there were bits of rhino hair tucked within the remaining bark... we were holding on to a tree that the rhino used as a place to scratch its back... fortunately and perhaps most unfortunate for the rhinos, the mating ritual was disrupted and the path they used to move far from human disturbance was taken in the opposite direction. (Personal communication. 1986. Anonymous travel Diary-Nepal)

The scenario described above demonstrates the influence a tour operation, mediated in this case by a tour guide, has both for the environment, the protected area, a species, and visitor experience. It also raises issues with how guides might perceive visitor expectations and many aspects of control, experience, and impacts. Further, it addresses the outcomes the guides on the ground manage—and the ripple effect, where tourism meets conservation or protected area, and how the result of these actions influence the community and perhaps even a larger regional context.

According to Fredericks et al. (2008) in today's society tour operators are seen as a "pivotal link between the tourist and the destinations, and thereby represent a leverage point for leading the move towards sustainability" (p. iv). As such, there is increased acknowledgement of the importance of tourism operator's ability to influence and shape the relationships between a destination's communities, natural and cultural heritage, and economic development, which represent the key outcomes of sustainable tourism (UNEP 2008). In this chapter we utilize the Tourism Satellite Accounting definition of a tour operator which states:

> tour operators are businesses that combine two or more travel services (e.g., transport, accommodation, meals, entertainment, sightseeing) and sell them through travel agencies or directly to final consumers as a single product (called a package tour) for a global price. The components of a package tour might be pre-established or can result from an "a la carte" procedure, in which the visitor decides the combination of services he/she wishes to acquire. (Tourism Satellite Account, cited in Organization for Economic Cooperation and Development 2008, p. 794)

Analogous to an ecosystem, "there is an existing understanding of the interdependency between the social, economic, and environmental systems: inherently high levels of cooperation between tour operators and their immediate stakeholders; the industry's movement towards sustainability, and the strategies, actions and tools that have arisen to reveal the link between sustainability and the business case for tour operators" (Fredericks et al. 2008, p. vii).

As such, tour operators are part of a larger system of providers which comprise three kinds of trades, including the *primary trades*, of which tour companies, transport, accommodations, attractions, catering, and travel agencies etc. are most commonly associated with tourism; the *secondary trades* that are supportive of tourism, though not exclusive to tourism, and include entities such as retail shops, banking, entertainment and leisure activities, insurance, etc.; and lastly, the *tertiary trades* which are the elements of society that provide basic infrastructure and support for tourism, which could include anything from public sector services to food, manufacturing and fuels (Likorish and Jenkins 1997). However, one of the challenges of exploring frameworks for tour operators in meeting the goals of sustainable tourism is the diverse range in the *scale* of tour operations, which span from international and national operators to locally based tour operators that deliver for example, whale watching tours. In this chapter we acknowledge these different levels of scale and attempt to identify ways that all tour operators at all levels can make a contribution to sustainability given the enormous environmental, social, and economic challenges faced in society today (See Chaps. 1 and 2; Orr 1994, 2011; UNEP 2008).

Recognized by industry leaders, the subsequent control of all aspects of the tourism product by tour operators is an increasingly challenging endeavor. As noted by the Tour Operators Initiative (TOI) (2004):

> because most of the goods and services included in a holiday package are provided by a supply chain of subcontracted companies, organizations and agents, tour operators are not always in direct control of the environmental and social impacts of those products. Yet, consumers increasingly expect the companies they buy from to ensure that their products provide not just quality and value-for-money, but also safeguard environmental and social sustainability (p. 3).

As mentioned above, the structure of the tourism industry is highly complex, and is continually evolving and changing with new technological advances and expanding ideas that further support globalization (Yeoman 2008). Despite this, it is tour operators specifically that are engaged in a broad range of responsibilities directly tied to a tourism destination. These relationships ultimately influence what occurs within the destination and the perception of the destination by those who visit it.

Fredericks et al. (2008) have demonstrated that the current model of a price driven market and short term thinking among tour operators needs reassessing to facilitate a comprehensive system, outcome focused, visionary approach to sustainability. Hence the role of the tour operator may be changing, and as part of a larger system, requires a new look at their function within the system and ultimately how they are managed. Therefore, the goal of this chapter is to identify the ever-evolving role of the tour operator in the sustainable development and management of tourism. We do this first through a distillation of project guidelines for tourism operators, personal experience, and our involvement in international organizations focused on sustainable tourism and ecotourism. Secondly, we utilize the Global Sustainable Tourism Criteria (GSTC 2014) as a basis for identifying the role of tour operators in managing sustainable tourism operations and their role at a destination level. We close with a proposal for a framework identifying the expanding role of tour operators in sustainable tourism management. This proposed framework is a culmination of years of work by many organizations, research projects, and industry leaders that have identified avenues to sustainable tourism management.

6.2 Tour Operators and Influence

In 2000, the Tour Operators Initiative for Sustainable Tourism Development, a voluntary non-profit organization was launched with the primary goal of advancing the sustainable development and management of tourism and secondly to make sustainability a main stream issue in tourism business (TOI 2003). Through TOI, there is recognition of the necessity for tourism operators to be an integral part of a sustainable society. According to the TOI (2008) "Tour operators are important for the destination's economy, and destinations are important for tour operators—without them, there would not be a tourism product." The TOI stated that the role of the tour operator within a destination includes:

- Influencing customers' choices and behaviors
- Directing the flow of tourists
- Influencing the supply chain
- Influencing the development of destinations
- Influencing the well-being of destinations and local communities

The amount of leverage a tour operator has to achieve more sustainable tourism is dependent on a range of factors such as the size, structure, and web of suppliers and services that they use and engage. Not only is size and scope of the tour operation of interest within these relationships, so is the nature of the complexity. For the sake of comparison, let's explore two adventure tour operators, one that exists within a small Pacific island nation and the other within North America. The relative influence of each depends not only on the size (i.e., number of clients served, number of employees) of the operation, but on the context and destination within which they operate. For example, with a relatively small operator in the Pacific, the tour operation is comprised of a day tour consisting of transportation, a lunch, and river experience; the tour operator in North America owns several permits within the Western United States, works with 1–20 day river experiences, and manages accommodations pre and post river trip, and sells packaged adventure tours overseas as well—hence, as an inbound and outbound adventure tour provider it influences several suppliers, and wields greater negotiation capability over its complex supply chain. In contrast, the adventure tour operator in the Pacific works directly with local communities to operate their programs on the rivers—hence, two similar product providers with very different influence and impact on the destination(s) where operations occur.

Important in the discussion is the "sphere of influence" tour operators potentially have with respect to these destinations where they operate (Budenau 2007). This influence is conceived at three levels varying from direct control to minimal control: (1) sphere of direct control (high), the way in which they operate internally and product development; (2) the sphere of influence (medium), would include areas of supply chain management and customer relations; and (3) the sphere of support (low) would include destination relationships (Budenau 2007; Tepelus 2005). While differences exist depending on size and reach of tour operators, large-size tour operators have demonstrated an ability to influence customer activities, supply changes, and the pace of destination development (Carey et al. 1997; Budenau 2005). Consider the potential influence of an inbound tour operator. Typically they would have "tour desks" set up within various lodging providers to cater to tourists interests. Each tour desk employee has the potential power to sell or not sell particular products within the country. It is only the discerning tourist that may move beyond the tour desk and seek a tour experience not on the sales sheet of the desk. The inbound tour operator also typically would set what type of commission is paid for selling the tour. In some cases, they utilize the commission structure to "highlight" or influence the sales of certain products over others. This of course varies by country, however the inbound tour operator has significant influence on where visitors go and what they experience as tourists—with a focus on sustainable tourism products or not.

The United Nations Environmental Program (UNEP) and the TOI provided guidance for tour operators to assist in moving sustainability forward within the tourism sector. Of the various business activities a tour operator would typically engage in, five action areas were identified, these included (UNEP and TOI 2005, p. 13):

1. *Internal Management*: this applies sustainability practices to the operations of the company's headquarters, and includes changes that are relative to energy efficiency, minimizing waste, and ensuring acceptable staff working conditions;
2. *Product Development and Management*: this applies to the design of tour programs and products that minimize negative environmental and social impacts, while producing acceptable economic returns;
3. *Supply Chain Management*: this applies to the selection and contracting of service providers and may include actions such as setting sustainability standards, assessing performance and supporting improvements of suppliers, or the provision of incentives for these improvements (e.g., preferred provider, increased commission, etc.);
4. *Customer Relations*: this applies to building consumer awareness of sustainability issues through information sharing and interpretation on appropriate visitor behavior, such as the purchase of local products, respecting local culture, and minimizing guest's negative environmental impact of their visit;
5. *Cooperation with Destinations*: this applies to the influence tour operators, either individually or through industry forums, can have on sustainability within the destination, including but not limited to destination planning decisions and engaging in activities to support the destination and society.

In addition to these areas of action, tour operators can also influence the volume of visitors entering destinations, and the type of demand for the destination through the way in which the destination is represented in marketing and promotion materials (Leiper 1990; Fredericks et al. 2008). This suggests a significant responsibility for tour operators when it comes to their role in working with a destination and the pace of change and sustainability of a destination (Fredericks et al. 2008; Tepelus 2005; Budenau 2005; Carey et al. 1997; Holden and Kealy 1996). Ultimately, these actions begin to set the stage on the role a tourism operator can play to achieve sustainability. The interconnectivity of the tour operator within a tourism destination system is demonstrated in Table 6.1 (TOI and CELB 2004). For example, tour operators can have a significant influence on a supply chain in terms of the service providers they select and contract with—i.e., whether they insist on suppliers meeting sustainability criteria, and whether they provide incentives to encourage more sustainable procedures and practices. For example Premier Tours, a US based tour operator specializing in safari tours in Africa, uses service suppliers and ground handlers who are committed to sustainable practices and who apply a responsible approach to tourism, conservation, and local community involvement. Their approach ensures a sustainable environment, local employment and other

Table 6.1 Tour operators potential primary contracted products and suppliers

Elements of tourism products	Suppliers
Accommodation	Hotels, bed & breakfasts, self-catering, serviced apartments, campsites, cruise ships
Catering and food and beverage	Restaurants and bars, grocery stores, farmers, fishermen, local commerce/markets, bakers, butchers, food wholesalers
Cultural and social events	Excursion and tour providers, sports and recreation facilities, shops and factories
Ground transport	Car rentals, boat rentals, fuel providers, gas stations, coach rentals
Ground services	Agents, handlers or inbound operators in the destination
Environmental, cultural and heritage resources of destinations	Public authorities, protected site managers, private concessionaires and owners
Transport to and from destinations	Public transport (e.g. trains), airports, scheduled air carriers, air charters, scheduled sea passages, chartered sea passages, coaches, cruises

Source: TOI and CELB (2004)

benefits to the local communities as well as a high quality tourism experience. Part of their approach entails Premier Tours monitoring its suppliers' performance (UNEP 2005). As another example, in Bonito, Brazil, tour operators work together to achieve a high level of environmental protection of sites and a quality ecotourism experience. Local tour operators formed an alliance with travel representatives who sell ecotourism experiences in the area. All representatives sell the tour for the same agreed upon price—so no price wars or competition based on pricing. In addition, there are voluntary agreed upon limits of how many people can join each tour. These limits are set and upheld by each operator to protect the environment, the natural experience, and safety of each guest. The entire industry in the area supports these limits and meets monthly to discuss aspects of their tours, what is happening in the industry, and guest satisfaction. In addition, some of the tour operators monitor the local fauna of the areas they operate in conjunction with scientists, conservation groups, and government requirements.

There are inherent and significant challenges to sustainable tourism development, and as such, challenges for the role of the tour operator in achieving this goal. The sheer size of the industry creates a concern over the availability of resources to support over one billion travelers roaming the world today and the increases projected in the future (UNWTO 2013). The absence of boundaries within the various businesses, public lands, and private lands that provide the places where tourism happens may inhibit an actual definition of influence of the industry. The tourism product itself, which is really a combination of components, creating an overall touristic experience or "macro-level" product, or a "place product" which occurs at a destination cannot be singularly managed. Many of these components are comprised of numerous individual services, all of which encompass both tangible (meals, bus seats, lodging) and intangible elements (climate, natural scenery, host relationships, etc.) purchased and expected throughout the overall experience (Berno and Bricker 2001). In addition, tourism products are perishable, are used by one or more persons at the same time, and are subject to external factors such as political unrest, natural disasters and change in demand that the tourism destination is unable to apply direct control (Berno and Bricker 2001).

Another challenge tour operations face is a lack of a common understanding of the practical application of sustainable tourism, including: limited recognition of standards that have been developed, low of credibility in sustainable tourism certification programs, and a lack of a critical mass of consumer's ability to identify or care about sustainable tourism products (GSTC 2014).

6.3 Codes of Ethics, Conduct, Guidelines and Certification in Sustainable Tourism

To assist in moving the tourism industry towards sustainability, organizations, governments, and tourism destinations have created voluntary codes that comprise "a set of rules" for a course of action to be followed, in this case steps towards sustainability (Black 2015). There are several examples of codes of conduct/guidelines that exist today. From the Antarctica Treaty Guidelines for Visitors to the Antarctic (Mason 2007) to The International Ecotourism Society's Code of Ethics for Ecotourism Operators (see www.ecotourism.org). While many of these are voluntary, there are instances where codes have an associated monitoring and reporting system (Black 2015). There has been proliferation of codes of conduct in sustainable tourism and ecotourism which apply to a range of stakeholders, including visitors, accommodation, and tour operations. While many are voluntary, some are also regulated by government entities, hence semi-formal, at one extreme, and voluntary with no checks and balances at the other extreme (Black 2015). Yet research has demonstrated that most codes of conduct are voluntary and problematic when it comes to monitoring the outcomes and adherence to them (Font and Buckley 2001; Issaverdis 2001). Because of the proliferation and rudimentary nature of these Codes, voluntary certification programs have been introduced to assist in verifying sustainable practices and moving the industry forward in sustainable action. Whether or not codes of conduct/ethics or certification are making a difference is still up for debate (Garrod and Fennell 2004)—with respect to compliance, what constitutes sustainability in tourism, and what metrics should be utilized. For example, in an analysis of 58 whalewatching codes of conduct Garrod and Fennell (2004) found there was considerable variability among the codes around the world, with respect to the voluntary nature and what was actually employed within the operation (Black 2015).

6.4 The Global Sustainable Tourism Council

To address these challenges, and bring unity to all of the guidelines and codes available today, the Global Sustainable Tourism Council (GSTC) was established as an international non-profit organization. The GSTC's primary objective is to bring together businesses, governments, non-governmental organizations, academia, individuals, and communities engaged in, and striving to achieve best practices to develop a common language in sustainable tourism. The GSTC compiles, reviews, adapts and develops tools and resources to foster sustainable tourism practices and increase demand for sustainable tourism products and services. It has engaged in the development of sustainable tourism criteria, which provides a foundation for many national programs and global policies establishing sustainable tourism development today (GSTC 2014). The criteria are a set of universal principles that create a common language and framework for sustainable travel for tour operators and accommodations, and most recently, for destinations.

The criteria are part of the response of the tourism community to the global challenges of the United Nations' Millennium Development Goals. Poverty alleviation, gender equity and environmental sustainability, including climate change, are the main cross-cutting issues that are addressed through the criteria. The GSTC Criteria and indicators were developed based on already recognized criteria and approaches including, for example, the UNWTO destination level indicators, GSTC Criteria for Hotels and Tour Operators, and other widely accepted principles and guidelines, certification criteria and indicators. They reflect certification standards, indicators, criteria, and best practices from different cultural and geo-political contexts around the world in tourism.

According to the GSTC, in order to achieve sustainable tourism a number of sectors of the tourism industry must be targeted—a multi-pronged approach—so criteria have been developed for destinations, tour operations, and accommodations (GSTC 2014). These criteria are all relevant to the operations of tour operators. For tour operations, accommodation, and destinations criteria take an interdisciplinary, holistic and integrative approach which includes four main objectives: to (i) demonstrate sustainable management; (ii) maximize social and economic benefits for the host community and minimize negative impacts; (iii) maximize benefits to communities, visitors and cultural heritage and minimize impacts; and (iv) maximize benefits to the environment and minimize negative impacts (GSTC 2014). The criteria are designed to be used by all types and scales of tour operations, accommodations, and destinations and to fully address some of the current issues arising out of unsustainable development. Some of the expected uses of the GSTC criteria by the tourism industry including tour operators are:

- Serve as basic guidelines for destinations which wish to become more sustainable;
- Help consumers identify sound sustainable tourism destinations;
- Serve as a common denominator for information media to recognize destinations and inform the public regarding their sustainability;

Table 6.2 Differences between tour operator and accommodation and destination criteria

GSTC criteria—tour operators and accommodation	GSTC criteria—destinations
Impacts under the company's control	Cumulative impacts of all activities in the destination
Specific impact mitigation actions	General impact mitigation actions
Benefits the immediate community	Involves the whole community as actors
Competitive advantage for the company	Competitive advantage for the destination and all of its businesses
Outreach to a tour company and its customers, employees and neighbors	Outreach to the community, tourism businesses, other businesses and local governments
Requires an involved management and trained employees	Requires one or more organizations as manager(s) of the destination

GSTC (2014)

- Help certification and other voluntary destination level programs ensure that their standards meet a broadly-accepted baseline;
- Offer governmental, non-governmental, and private sector programs a starting point for developing sustainable tourism requirements; and
- Serve as basic guidelines for education and training bodies, such as hotel schools and universities (GSTC 2014).

The GSTC has developed two sets of criteria with different emphases. The first is tour operators and accommodation focused, and the second is destination focused. All these criteria effect and influence tour operators in different ways, and at different scales. Table 6.2 summarizes the primary differences between the Criteria for tour operator and accommodation sectors and criteria for destinations. For more information on the development of both sets of Criteria and supporting indicators, see www.gstcouncil.org.

The GSTC Criteria were conceived as the beginning of a process to make sustainability the standard practice for all forms of tourism, as well as define sustainable tourism in a way that is actionable, measurable and credible. The Criteria are envisioned as the minimum standard of sustainability for tourism businesses and destinations across the globe (GSTC 2014). The criteria assist the industry in re-establishing and re-thinking the role of tour operators in implementing sustainable tourism development and management.

6.4.1 Application of Tour Operator and Accommodation Sector Criteria to ST

In an effort to link the role of the tour operator to sustainable development and management of tourism, the criteria indicate what should be done, but not what actions are required to meet the criteria nor whether a goal has been achieved. The indicators are recommendations about ways of measuring compliance with the criteria (GSTC 2009, p. 2).

The following sections highlight some ways that tour operators can help meet the GSTC criteria through their internal and external operations, business model, staff, interactions with destination and approach to sustainability.

6.4.2 Demonstrating Effective Sustainable Management

6.4.2.1 Tour Operational Level

To demonstrate the company is managed sustainably a tour operator must consider and assess their business model and the daily company operations. For example, they need to take an active role in ensuring that all legal and regulatory requirements are met by the company and relevant staff. This means that staff need to be trained to ensure this goal is achieved as well as having opportunities for professional development. Customers would have the opportunity to evaluate their experience or register comments on the experience. In terms of the tourist experience, locally based tour operators have a key part to play in meeting sustainability goals particularly through the role of the employees that have direct interaction with visitors, such as the tour guide. Guides can play in important role in fostering and achieving sustainable tourism outcomes (Wieler and Kim 2011). For example, Weiler and Kim identified four dimensions of tour guiding practice that can be used to better harness tour guides as agents of ST outcomes. Wieler and Kim (2011) contend that tour guides can contribute to sustainability by:

1. *Enhancing* visitors' *understanding and valuing* of the site and its natural and cultural resources, *through interpretive guiding*;
2. *Influencing* visitors' decisions about their voluntary onsite behaviour, *through communicating and role modelling* ST practices;
3. *Monitoring and managing* visitors' *on-site behavioural compliance, through enforcing regulations and role modelling practices* associated with protecting ecological and cultural values; and
4. *Fostering* visitors' *post-visit* pro-environmental and pro-conservation *attitudes and behaviours, through persuasive communication* (p. 114).

Clearly through these roles the guide can achieve many sustainability outcomes of education and awareness and supporting protected area managers in enforcing regulations and contributing to environmental protection (Armstrong and Weiler 2002). The role of the tour operator here is to employ well trained, passionate guides, to provide good wages and conditions that encourage professionalism among the guides, and continuous professional development, which may include certification (Black and Weiler 2005).

With respect to site planning, if a tour operator is involved in the planning and design of a new facility they would ensure that they comply with local planning laws and respect the natural and cultural environment in siting, design, impact assessment and land rights acquisition. See for example Ecolodge design and planning (Mehta 2007, 2013).

6.4.2.2 Tour Operator Involvement at the Destination Level

At the destination level, a tour operator would likely be involved in the organization and coordination of tourism, and participate in a monitoring program related to environmental, economic, social-cultural, and potentially human rights issues. They would also participate in the collection of data with regards to tourist satisfaction, they would take part in safety and security planning at the destination level, participate in a crisis and emergency response plan, and ensure that promotional materials are accurate in their description of products and services at the destination level.

6.4.3 Maximize Social and Economic Benefits to the Local Community and Minimize Negative Impacts

6.4.3.1 Tour Operational Level

The role of a tour operator can also contribute by bringing economic benefits through the employment of local residents (including management positions) and by offering ongoing professional development opportunities for its staff. The operator would have policies against social injustices such as commercial exploitation of children and adolescents, including sexual exploitation. In addition, a tour operator would actively support initiatives for social and infrastructure community development in areas such as health, education, and sanitation. The tour operator may also support and use the services of local small entrepreneurs to develop and sell sustainable products based on the local nature, history and culture such as food and drink, crafts and performance arts.

6.4.3.2 Tour Operator Involvement at the Destination Level

At the destination level, tour operators would once again take participatory action within their community. For example, they would participate in a localized system which involves stakeholders in destination planning and decision making. They would participate in programs designed to raise awareness of tourism's role and potential contribution in economic development in their area. They would implement ways to contribute to sustainable community and biodiversity conservation initiatives. They would support local purchases and utilize local suppliers in the tourism value chain such as artisans and farmers.

6.4.4 *Maximize Benefits to Cultural Heritage and Minimize Negative Impacts*

6.4.4.1 Tour Operational Level

Tour operators can ensure the company and its tour guides follow visitor and operator's guidelines or codes of conduct when visiting culturally sensitive sites in terms of minimizing visitor impacts and maximizing tourist satisfaction. They would ideally participate in implementing codes of best practice applicable to their guiding strategies and visitor management techniques. Further, they would incorporate elements of local art, architecture, or cultural heritage in their operations, design, decoration, foods, or shop, all the while respecting the intellectual property rights of local residents and/or communities.

6.4.4.2 Tour Operator Involvement at the Destination Level

Tour operator participation at the destination level would entail visitor management plans, respecting and implementing guidelines and codes of practice, and being a vehicle or tool for implementing interpretation services for visitors.

6.4.5 *Maximize Benefits to the Environment and Minimize Negative Impacts*

6.4.5.1 Tour Operational Level

Tour operators can meet this goal in many ways. Operationally, they could monitor their use of energy in their operations and by purchasing sustainable disposable and consumable items and through actively seeking ways to reduce their use. They may also support conservation efforts through financially or through in-kind services. Because tour operators are dependent often times on the natural or cultural capital as their primary attraction, they may also be directly involved in managing or protecting the experience through visitor use limits, low impact use strategies, and working with land managers on monitoring impacts over time.

The tour operator can carry out a number of measures to meet the goal of reducing pollution and waste. For example measuring the greenhouse gas emissions for all the sources controlled by the operation and implementing procedures to reduce and offset them as a way to achieve climate neutrality.

Finally, a tour operator can assist in conserving biodiversity, ecosystems, and landscapes by supporting biodiversity conservation, including supporting protected areas and protecting areas of high biodiversity value. This can be achieved through the company or tour guides encouraging tourists to donate time or money to specific local development projects (Ham 2001; Powell and Ham 2008; O'Brien and Ham

2012). In other instances this is done through direct conservation supplements, such as developing a lease for conservation (Rivers Fiji 2014), or through partnership agreements with protected area lands or marine areas to enhance or restore through on-going voluntary initiatives.

6.4.5.2 Tour Operator Involvement at the Destination Level

The tour operator at the destination level participates in strategies to support management systems in place to prevent causing negative environmental impacts. This may include items such as participating in programs that monitor, minimize, and report energy use, water use, water quality issues, solid waste reduction, noise and light pollution. The responsibility of the tourism operator is again active and participatory, and moves stewardship of the destination forward.

In summary, tour operators can influence the sustainability of destinations through complex decisions that affect the environmental, social and economic sustainability of the destination. For example, by using their business influence to create a long-term vision for a destination by working in partnership with local stakeholders to create a high quality tourist experience that protects the destination's economy, culture and environment as well as increasing benefits to the local community (UNEP 2005). Local partners may include other tour operators, protected area managers, NGOs, government authorities and of course local residents.

6.5 Summary Framework for Sustainable Tourism Operators

In considering the wide variation in tour operators and the products they provide, we have developed a framework which captures the variability and potential role and influence tour operations have within the industry overall. This framework (Table 6.3) provides guidance when considering tourism operators within as part of a sustainable tourism industry, and the diversity which occurs at varying levels of influence and scope. Within our framework we address the type of the tourism operation (i.e., global to local), the scale and hence, area of influence tourism operators have (Budenau 2005), the various actors or stakeholders involved, the opportunities to influence or impact sustainable tourism, and the benefits or outcomes of these actions.

The scale of the tourism operation reflects whether or not the tour operator is operating within an international context or locally. This forms the basis on how widespread and/or the level of control the operator has in initiating sustainability overall. The scale or area of influence then determines what types of opportunities there are with various stakeholders involved in producing tourism products, and therefore what benefits or impacts these have on sustainability within the sphere of influence of that particular operator.

Table 6.3 Summary framework for sustainable tourism operators

Scale of tour operation	Scale/area of influence	Actors	Opportunities to influence/impact sustainable tourism	Benefits/outcome/impact on sustainable tourism
International tour operation (outbound)	International, national, regional, local	Tour operators and staff	Supply chain management	Local suppliers used
		Governments	Educate tourists pre departure	More informed and educated tourists
		Suppliers—accommodation, hospitality, transport, etc.	Employ local people	Benefit to local economies—multiplier effect
		Tourists	Internal management and operations	More sustainable businesses
		Travel agents	Product development, diversity and management	More sustainable tourism destinations and tourism industry—internationally, nationally and locally
			Destination planning and cooperation	
National and regional (inbound)	National, regional, local	Tour operators and staff	Supply chain management	Local suppliers used
		Governments	Educate tourists pre departure	More informed and educated tourists
		Regional associations and governments	Employ local people	Benefit to local economies—multiplier effect
		Suppliers—accommodation, hospitality, transport, etc.	Internal management and operations	More sustainable businesses
		Tourists	Product development, diversity and management	More sustainable destinations and tourism industry—nationally, regionally, locally
		Protected area managers	Destination planning and cooperation	

Local	Regional and local	Tour operators	Supply chain management	Local suppliers used
		Governments	Employ local people	Tourist's increased understanding and value of the natural and cultural environments
		Local governments	Provide interpretation to tourists	Fostering visitors' post-visit pro-environmental and proconservation attitudes and behaviours
		Local suppliers—accommodation, hospitality, transport, etc.	Guides role model appropriate environmental and cultural behaviours	Benefit to local economies—multiplier effect
		Tourists	Guides mediate experiences with local communities and cultures	Communities directly benefit from tourism—employment, handicraft sales, cultural displays and events
		Guides	Liaise and work with local communities	Traveller philanthropy—schools, scholarships, health centres, volunteerism
		Indigenous and non-Indigenous Communities	Internal management and operations	Monitoring and managing visitors on-site behaviour
		Protected area managers	Product development, diversity and management	Minimal impact messages assist in protecting natural and cultural environments
			Destination planning and cooperation	More sustainable businesses
				More sustainable destination and tourism industry locally
				Support for protected area managers

Imagine a company like Tui Travel, one of the largest tour companies in the world, serving over 30,000 hotels and accommodations alone—as an international corporation, they have a large and complex system. Their influence is enormous, with many varied opportunities for influence and ultimately large scale benefits covering a large geographic area. This framework provides an opportunity to see where the actions of tour operators influence supply and how the varied types of operators influence sustainability, from a global to a localized level of action.

6.6 Summary and Conclusions

In daily life, regions and businesses are interlinked systems into a web of people and nature, driven and dominated by the manner in which they respond to and interact with each other, that is they are complex systems continually adapting to change. According to Walker and Salt (2006) "The ruling paradigm that we can optimize components of a system in isolation of the rest of the system- is proving inadequate to deal with the dynamic complexity of daily life. Sustainable solutions to our growing resource problems need to look beyond a business as usual approach" (p. 8). Due to the size and reach of the tourism sector, its enormous growth rates and risks of negative impacts, sustainable development and management of the tourism system is of critical importance, in order to achieve a balance between a healthy ecosystem, quality of life, high satisfaction of tourists, while balancing long-term business success.

Ultimately, sustainable tourism is not a special form of tourism or a product, but strives to sustain all elements of the tourism system, not simply the business. Research has demonstrated that tour operators can contribute to creating more resilient communities and destinations (GSTC 2014; Fredericks et al. 2008) and they realize they can contribute to sustainability and are becoming more proactive.

Tour operators recognize the need, as well as increased demand from tourists, communities and governments for longer term relationships that benefit social, economic and environmental sustainability. With the advent of the GSTC criteria and other global initiatives focused on tourism operators such as the TOI, there is recognition of the need for a shared vision and common understanding of sustainability in the tourism sector. Fredericks et al. (2008) concluded that: "tour operators need a more comprehensive planning framework when looking at sustainability. This requires a whole-systems perspective, in order to address the complexities inherent in the industry and to ensure the survival of the tour operator's business and the destinations they represent" (p. 58). This approach includes the active participation of stakeholders within a destination and includes a governance/policy/planning dimension, economic dimension, social-cultural dimension, and environmental dimension. This systematic approach is impacted by the type and influence of the tour operator (see Table 6.3).

Several tools are available to understand how tourism operators may implement and influence sustainability within the destinations they work. First, the Global Sustainable Tourism Criteria and associated indicators provide the baseline guidance for sustainable tourism (see gstcouncil.org). Secondly, organization such as United Nations Environment Programme, and the International Labour Organization have toolkits that assist tourism operators in implementing sustainable tourism (see http://www.unep.org/ and http://www.ilo.org/global/lang-en/index.htm). These organizations have published toolkits, best practices, and case studies which can assist stakeholders in understanding sustainable tourism implementation strategies.

And finally, the role of the tourism operator is changing and becoming more complex as we learn of the potential influence they have at varying operational levels as they are part of a complex system of suppliers and product developers. As tourism numbers increase, greater demand is placed on all resources within the supply and value chain, largely promoted and utilized by tour operators. In a world of dwindling resources, tour operators must look at how they are valuing ecosystem services, because inherent in the promise of sustainable tourism as David Orr (1994) suggests is that we take a closer look at the way in which we "do" business. He suggests that we need to learn "how to build local prosperity without ruining some other place... And, to revitalize an ecological concept of citizenship rooted in the understanding that activities that waste resources, pollute, destroy biological diversity, and degrade the beauty and the integrity of the landscape are forms of theft from the common wealth..." (p. 168). Tourism operators are the link between the destination and the tourist and as such play an important role in moving the tourism industry to a more sustainable future.

References

Armstrong, E.K., and B. Weiler. 2002. Getting the message across: An analysis of messages delivered by tour operators in protected areas. *Journal of Ecotourism* 1(2): 104–121.

Berno, T., and K. Bricker. 2001. Sustainable tourism development: The long road from theory to practice. *International Journal of Economic Development* 3(3): 1–18.

Black, R.S. 2015. Codes of conduct. In *Encyclopaedia of sustainable tourism*, ed. B. Garrod and C. Cater. Wallingford: CABI Publishing.

Black, R., and B. Weiler. 2005. Quality assurance and regulatory mechanisms in the tour guiding industry: A systematic review. *Journal of Tourism Studies* 16(1): 24–37.

Budenau, A. 2005. Impacts and responsibilities for sustainable tourism: A tour operator's perspective. *Journal of Cleaner Production* 13: 89–97.

Budenau, A. 2007. *Facilitating transitions to sustainable tourism-the role of the tour operator*. Doctoral dissertation, IIIEE Lund University.

Carey, S., Y. Gountas, and D. Glbert. 1997. Tour operators and destination sustainability. *Tourism Management* 18(7): 425–431.

Font, X., and R. Buckley (eds.). 2001. *Tourism ecolabelling: Certification and promotion of sustainable management*. Wallingford: CABI.

Fredericks, L., R. Garstea, and S. Monforte. 2008. *Sustainable tourism destinations: A pathway for tour operators*. Master's thesis, School of Engineering, Blekinge Institute of Technology, Karlskrona.

Garrod, B., and D. Fennell. 2004. An analysis of whalewatching codes of conduct. *Annals of Tourism Research* 31(2): 334–352.

Global Sustainable Tourism Council. 2014. The global sustainable tourism criteria. www.gstcouncil.org

GSTC. 2009. The Global Sustainable Tourism Council Criteria (v1) abbreviated indicators 2009. http://www.gstcouncil.org/resource-center/progress-indicators.html

Ham, S. 2001. A theory-based approach to campaign planning for travelers' philanthropy. Invited address to Traveler's Philanthropy Summit, Hosted by Business Enterprises for Sustainable Tourism (BEST). Punta Cana Resort, 9–11 Nov 2001.

Holden, A., and H. Kealy. 1996. A profile of UK outbound "environmentally friendly" tour operators. *Tourism Management* 17(1): 60–64.

Issaverdis, J.-P. 2001. The pursuit of excellence: Benchmarking, accreditation, best practice and auditing. In *The encyclopedia of ecotourism*, ed. D. Weaver, 579–594. Wallingford: CABI.

Leiper, N. 1990. *Tourism systems*, 1–40. Palmerston North: Department of Management Systems, Massey University.

Likorish, L.J., and C.L. Jenkins. 1997. *An introduction to tourism.* Oxford: Butterworth-Heinemann.

Mehta, H. 2007. Towards an internationally recognized ecolodge certification. In *Quality assurance and certification in ecotourism*, ed. R. Black and A. Crabtree. Wallingford: CABI International.

Mason, P. 2007. No better than a band-aid for a bullet wound: The effectiveness of tourism codes of conduct. In *Quality assurance and certification in ecotourism*, ed. R.S. Black and A. Crabtree. Wallingford: CABI.

Mehta, H. 2013. Small island protected area planning: A case for West Caicos. In *Sustainable tourism and the millennium development goals: Effecting positive change*, ed. K.S. Bricker, R. Black, and S. Cottrell. Burlington: Jones and Bartlett Learning.

O'Brien, T., and S.H. Ham. 2012. *Toward professionalism in tour guiding: A manual for trainers.* Washington, DC: US Agency for International Development.

Organization for Economic Cooperation and Development. 2008. Cited in Tourism Satellite Account (2001) *Recommended methodological framework.* Eurostat, OECD, WTO, UNSD, para 3.46. Paris, France.

Orr, D.W. 1994. *Earth in mind.* Washington, DC: Island Press.

Orr, D.W. 2011. *Hope is an imperative.* Washington, DC: Island Press.

Powell, R.B., and S.H. Ham. 2008. "Can ecotourism interpretation really lead to pro-conservation knowledge, attitudes and behaviour?" Evidence from the Galapagos Islands. *Journal of Sustainable Tourism* 16(4): 467–489.

Rivers Fiji. 2014. Upper Navua conservation area. http://www.riversfiji.com/ecotourism

Tepelus, C.M. 2005. Aiming for sustainability in the tour operating business. *Journal of Cleaner Production* 13: 99–107.

Tour Operators Initiative. 2003. *Sustainable tourism, the tour operators.* Paris: UNEP.

Tour Operators Initiative (TOI). 2008. About TOI. http://www.toinitiative.org/index.php?id=3

Tourism Operators Initiative and Center for Environmental Leadership for Business (CELB). 2004. *Supply chain engagement for tour operators: Three steps towards sustainability.* Tour operators initiative for sustainable development, Madrid. http://www.toinitiative.org/fileadmin/docs/publications/SupplyChainEngagement.pdf

Tourism Satellite Account (2008). Recommended Methodological Framework. Commission of the European Communities, Eurostat, Organisation for Economic Co-operation and Development, World Tourism Organization United Nations Statistics Division, Luxembourg, Madrid, New York, Paris, 2008

United Nations Environment Program (UNEP). 2008. Tourism impacts. http://www.unep.fr/scp/tourism/sustain/impacts/

United Nations Environmental Program and the Tour Operators Initiative (UNEP and TOI). 2005. Integrating sustainability into business: Management guide for responsible tour operations. http://www.unep.fr/scp/publications/details.asp?id=DTI/0690/PA

United Nations World Tourism Organization(UNWTO). 2013. UNWTO tourism highlights 2013 edition. http://mkt.unwto.org/publication/unwto-tourism-highlights-2013-edition

Walker, B., and D. Salt. 2006. *Resilience thinking: Sustaining ecosystems and people in a changing world*. Washington, DC: Island Press.

Wieler, B., and A.K. Kim. 2011. Tour guides as agents of sustainability: Rhetoric, reality and implications for research. *Tourism Recreation Research* 36(2): 113–125.

Yeoman, I. 2008. *Tomorrow's tourist: Scenarios & trends*, Advances in tourism research series, 1st ed. Amsterdam: Butterworth-Heinemann, Elsevier.

Chapter 7
Tourism in Protected Areas: Frameworks for Working Through the Challenges in an Era of Change, Complexity and Uncertainty

Stephen F. McCool

Abstract Recreation planning and management is confronted with a growing series of challenges emanating from fundamental shifts in social values and preferences, large scale demographic changes, development of new technologies, economic restructuring and new perspectives on protected area governance and decision-making. These driving forces have led to increased complexity and uncertainty for protected area visitor and tourism management situations that can only be described as contentious and uncertain, with growing public scrutiny and demands for greater accountability. Combined with growing and diversifying demand for tourism, the stakes involved in protected area management have increased dramatically. This chapter argues that these forces have led to a situation where the intellectual capital needed for visitor management decisions has risen, the requirement for frameworks to help managers "work through" complicated decisions has escalated, and the need for understanding the consequences of decisions to recreation opportunities has grown. Against a backdrop of contentious decision-making and changing definitions of sustainable tourism, what frameworks are available for working through decisions and how suitable are they?

Keywords Recreation planning • Planning framework • Protected area tourism • Decision-making

7.1 Introduction

Protected areas play increasingly important roles in providing settings for recreation experiences and tourism development. While these areas have long afforded opportunities for resource commodities, such as timber, grass and minerals to supply local industries, while they often have served as the catchments for community water supplies, and while they continue to function as habitat for a variety of plants and

S.F. McCool (✉)
University of Montana, Missoula, MT, USA
e-mail: Steve.McCool@cfc.umt.edu

S.F. McCool, K. Bosak (eds.), *Reframing Sustainable Tourism*,
Environmental Challenges and Solutions 2, DOI 10.1007/978-94-017-7209-9_7

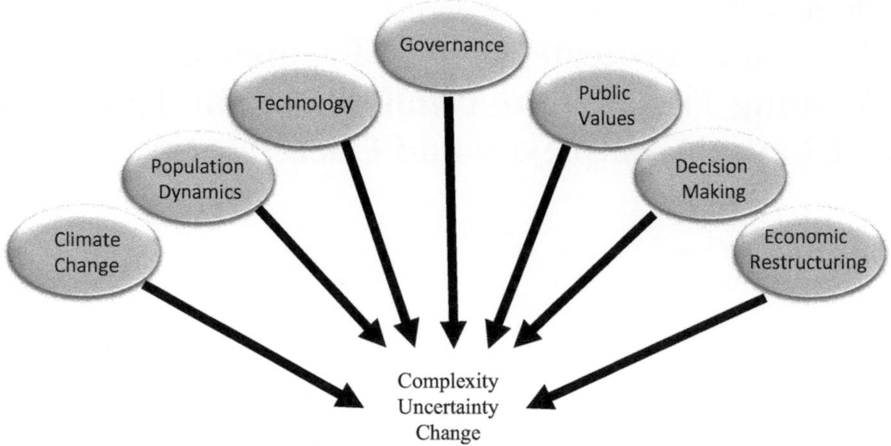

Fig. 7.1 A number of global level processes and forces affect the practice of sustainable tourism in protected areas, leading to complexity, uncertainty and change

animals, the search for sustainable tourism as a goal puts new responsibilities and demands on decision-makers. Tourism and recreation operate in somewhat different ways than resource commodities and traditional economic sectors; aesthetics is an important component, but so is good habitat, clean air and pure water. Private developments that exploit protected areas for recreation frequently occur off these lands, in communities and on adjacent parcels of land.

A variety of driving forces and contextual changes (e.g., population growth, changing public values, evolving philosophies about governance—see Fig. 7.1) have led to an increasingly contentious decision-making environment for recreation and tourism development, to decisions that are more complex, to situations where the stakes are higher, and to growing scrutiny and accountability in protected area planning. Overlaid upon this messy situation is a rising lack of trust in the capability of the government to make decisions in the interest of the public it is supposed to serve. These characteristics suggest that analysis of proposals, strategic policy, and project planning must rely more upon frameworks and concepts that explicate decisions than in the past, partly to avoid unnecessary impacts, duplication and loss of opportunities, and partly to ensure the optimization of benefits flowing from public lands.

And while conceptual advances in land management, such as landscape ecology, coupled with technological improvements, such as GIS, provide a greater capability for *informing* decisions, these changes have often lead to greater visibility of scientific uncertainty in those decisions. In a very real sense, as science has given us more knowledge it has also made our ignorance more visible. The consideration of longer time frames and larger spatial scales in decision analysis-resulting from increased attention to notions of sustainability—means we know less than before, and that there is greater argument over what we think we know. Too, the recognition embedded within such fields as landscape ecology that systems tend to be nonlinearly dynamic brings a certain ambiguity to the decision-making environment. These have converged to make it difficult for the public, to say nothing of the agency staff, to understand how decisions are made at precisely the moment in time when the public is demanding greater involvement in such decisions.

Recreation and tourism development are not immune from this situation. What once was perceived as a relatively "benign" use of protected areas is often as controversial as the resource extraction it has replaced. For example, in Yellowstone National Park, providing opportunities for winter recreation through snowmobiling has led to a variety of problematic social effects in addition to environmental ones. The result has been a decades long controversy with at least seven revisions in the Park's winter use plan, leading Borrie and others (2002) to conclude "that the need for an understanding of human values results partly from a limit to technical solutions for park management issues". This conclusion is similar to Dustin and Schneider's (2005) statement that the snowmobiling controversy is as much about the purpose of the Park as the consequences to wildlife.

Coupled with growing contentiousness over the values of protected areas, decision-makers providing opportunities for recreation and tourism development are confronted with increasing complexity and uncertainty about the consequences of their decisions. This occurs within a political climate that has raised the stakes involved in decisions, and with the need to scientifically justify plans and policies. All this arises within a context of accelerating change and connectedness, leading to linkages time zones away from the point of the decision

The result of this socially and scientifically turbulent environment is a growing need for frameworks and concepts that assist decision makers in assembling a set of informed alternatives and evaluating them. Concepts that are useful to decision makers are ones that help clarify conflict and opportunity and build understanding of choices and consequences. Useful frameworks are those that help decision makers "work through" these choices in a manner that allows technical expertise, knowledge (of various forms) and public values and interests to be incorporated, assessed and used.

In this chapter, I outline some of the issues and questions associated with the use of planning frameworks to resolve recreation and tourism management challenges in protected areas in this context. The overall purpose is to create a better understanding of why the use of a framework to work through decisions is so important, particularly in a time of change, complexity and uncertainty. Within a context of reframing sustainable tourism, critical thinking skills are essential to management. In addition, I discuss some of criteria to consider when adopting a particular framework.

7.2 Protected Areas and Recreation Planning Frameworks

The provision of recreation and tourism opportunities in protected areas within a dynamic, multidimensional and uncertain context is complex, challenging and fraught with potential misdirection and at the least surprises and unanticipated consequences. For example, increased demand that public lands provide commercialized recreation opportunities have led to conceptualizing management as one of identifying a carrying capacity (sometimes referred to as a "visitor capacity" for recreation, and then allocating such capacity between commercial and public visitors. Such a simplistic representation of a complex problem (e.g., what opportunities to be provided, to whom, where, how and with what consequences) follows from a lack of capacity needed to properly frame and respond to the problem. Lack of capacity to deal with managing recreation and tourism in protected areas is a globally significant question (McCool and others 2012) requiring new conceptual and practical skills. As in other areas of resource management, ideologies may subtlety influence the approaches managers take to such decisions about recreation and tourism.

Development of recreation and tourism frameworks has occurred in just the last 35 years. In the U.S. such development was generally in response to specific planning and implementation issues, often derived out of formalized policy. For example, in the U.S., the National Forest Management Act of 1976 and the National Park Service General Authorities Act of 1978 require better attention to recreation in planning and management. Globally, the Aichi Targets resulting from the Convention on Biodiversity require more areal coverage of the earth in the protected areas which means greater attention to managing them to include tourism and recreation.

However, these frameworks were also often initiated in response to opportunities and challenges that are somewhat different or evolved out of attempts to use a framework in situations for which it was not suitable. As Nilsen and Grant (1998) argue, the first step in determining suitability of frameworks is to "decide which questions they are seeking to answer". For example, the Limits of Acceptable Change framework (Stankey et al. 1985) was developed in response to numerous failed attempts to establish recreational carrying capacities for components of the U.S. National Wilderness Preservation System (McCool and others 2007).

A limited number of frameworks in this arena exist, and many have similar characteristics, but may have been developed in specific policy and administrative contexts that influences the particular elements or components involved. We note that several overviews and comparative analyses of recreation and tourism frameworks exist (Manning 2004; McCool and others 2007; Nilsen and Grant 1998; Moore et al. 2003). Each is helpful in familiarizing the reader with these frameworks; however, these reviews are generally not directed toward understanding their usefulness in addressing the variety of issues confronting managers. These reviews also tend to focus on a narrow set of frameworks, primarily around the Recreation Opportunity Spectrum (and its derivatives) and Limits of Acceptable Change processes.

In this chapter, I take a somewhat different course, laying out a functional description of what a recreation planning framework is, describing the criteria that would be needed to evaluate the suitability of a framework in a given situation, and proposing what conditions are needed to use a framework. While I mention the actual frameworks, they are not described here. The paper concludes with some observations about the future, the role of organizational capacity, and the some observations about how such frameworks are diffused through a bureaucratic organization.

There may be as many definitions of planning as there are planners, but probably the most wide spread approach to a definition of planning is that it is a process to both describe a desired and/or acceptable future and the "best" route to it—leaving open the definition of best. While other definitions range from "application of science to policy" to "linking knowledge to action" the one used here—see below—is probably the most widespread notion of the idea of planning.

Organizations plan for a number of reasons: to solve a problem, because they are told to do so, to reduce administrative discretion, to maintain consistency, to control that which can be controlled and so on. These notions may have been useful in the days of stability and predictability (if there ever were those days), but in an era that is chaotic, dynamic and filled with uncertainty, such concepts of planning don't seem to fit well. There are at least three weaknesses with those approaches to the idea of planning: (1) plans are built from assumptions, and therefore contain expectations about the future; those expectations, however, filter out important, contradictory information such that data challenging the validity of those assumptions may never be observed; (2) plans are contingencies, based on what we expect to occur, and thus limit our repertoire of potential actions should our expectations not be met and things change; and (3) planning processes presume that rational people, following the same process will come to the same decisions.

These assumptions are at odds with a world full of surprises, change and complexity (Weick and Sutcliff 2001). Planning in such a world must differ from the past. Kohl and McCool (2016) argue that our planning processes are direct descendants of world views and mental models, which often were developed in a world of modernity and post modern critique. Such foundational filters obscure reality while encouraging us to deal with symptoms rather than underlying causes. Kohl and McCool argue that our planning needs to be more holistic as well as oriented toward implementation rather than producing documents, which we know as plans.

Planning then becomes a way of thinking critically. An example is provided by Nkhata and McCool (2012) where they propose protected area planning as a kind of coupling activity joining governance and management:

> We believe that successful planning effectively couples governance and management subsystems for timely and appropriate responses to societal demands, provides for the experimentation needed for effective operations and learning in dynamic contexts, and encourages the assessment, reflection and learning underlying such activities. Viewing planning as a collective, iterative and adaptive process of creating and facilitating opportunities for dialogue leads to building and sustaining of the shared understandings within and across governance and management subsystems that are needed for societal action.

The linear, unidirectional character of conventional planning then is not well-suited for the contexts within which recreation and tourism management often occurs. These settings are not only contentious, but are fluid as well, with shifting priorities, changing needs and evolving opportunities and challenges. Government bureaucracies are established in part to address routine and repetitive problems, those with well-tested procedures and agreement on goals, commonly known as tame problems. Contemporary sustainable tourism challenges, such as exemplified in Table 7.1, represent conflicting goals, competing uses, and divergent views on tourism in protected areas with few known or tested processes for addressing them. These situations represent messy situations.

Development of planning frameworks can be viewed as an evolving critique of the inadequacies of government procedures to address complex problems in contentious situations. Messy situations not only require systematic processes that explicate fundamental assumptions and perspectives but also those that incorporate differing value systems and types of knowledge. In this sense, planning can be viewed as an iterative, inclusive process where constituencies and planners jointly frame issues, construct futures, and choose socially acceptable, efficient, equitable and effective pathways to those futures.

Given the above, frameworks that focus on allocation decisions for tourism serve to provide a systematic process in making those decisions such that managers are fully aware of the desired future they wish to attain, the alternative routes to the future and the consequences of those alternatives. In addition, these frameworks provide the explicitness and feedback needed in a time of change, complexity and uncertainty.

Finally, recreation and tourism planning frameworks make decision-making more efficient by focusing attention to important elements of the political and social environment, more effective by gaining the public support that is needed for implementation, and more equitable by forcing consideration of who wins and who loses. In an overall sense, a framework increases the opportunities to practice the "mindfulness" Weick and Sutcliff (2001) argue that is important to deal with the inevitable surprises occurring in an uncertain context.

7.3 What Is a Planning Framework?

A "framework" may be defined as a process involving a sequence of steps, a set of questions or components or even a diagram that leads managers and planners to explicate the particular issue. A "framework" in this sense does not necessarily lead to formulation of "the" answer to an issue, but provides the conceptual basis through which thinking is conducted, questions asked or discussion occurs leading to resolution of the issue. Frameworks are structures that enable us to apply critical thinking skills to a complex problem, they are not processes that can be simply followed without understanding their underlying rationale and conceptual underpinnings. Frameworks allow us to create deeper understanding of these issues

Table 7.1 Example issues confronting protected area tourism and recreation managers in the twenty-first century

Issue	Description
What are the interactions between tourism and other uses of protected areas?	Tourism and visitation frequently occurs within the context of other uses, both utilitarian, such as timber harvesting, grazing, habitat protection, and symbolic, such as visual quality, spiritual meanings. How visitation is managed affects the ability of protected areas to produce or preserve these other uses, and conversely, management for these other values influences what opportunities, where and how many there exist for recreation. In allocating lands to various uses, planners need to understand what trade offs, costs and consequences result from different proposed allocation decisions
Under what conditions can visitation be limited, and what criteria would be needed to make visitor use allocation decisions?	In certain situations, managers may feel that visitor use must be limited to a certain number of people during a specific time period. Such use limits have often been implemented on western whitewater rivers in the U.S. Use limits are generally implemented when there have been clear threats to the biophysical or experiential component of a particular setting. When use limits are imposed and demand is above what the limit allows, use must be rationed and allocated. By allocation, we mean dividing up the total use among commercial outfitted groups and private visitor groups, a common practice in river situations. Rationing is the process for determining within each of these groups the specific individuals that are permitted to enter the setting
What are the regional level effects of site level decisions?	Tourism sites, and larger areas, such as national parks, exist within a complex web of interacting supply and demand processes. Managers acting to protect or enhance the recreation attributes or tourism opportunities at one site may implement a series of actions that restricts people or their behavior, but in reality, the problem pops up someplace else. For example, acting to limit use on one site may displace use to other sites because at least some users can no longer access the original site. In some situations, the organizational capacity to deal with increased use and impact may be very limited on the sites where use is displaced to
How can allocations of use opportunities between outfitted and non-outfitted publics be made?	Limiting visitor use will often require that limits also be placed on the number of "service days" of use allowed for outfitted use (tour operators) and on the number of visitor days of use allowed for the non-outfitted public. This decision is akin to cutting a highly desirable pie into two slices (sometimes more, depending upon agency and outfitting policy). Criteria are needed to determine what proportion of use should be outfitted and non-outfitted. Within this decision there are also decisions to allot use to individual outfitters and to ration use among non-outfitted visitors, assuming demand is above the allocated use. Outfitters generally ration use based on price, while for the non-outfitted visitor, rationing may be based on a waiting line, reservation, random drawing or a combination of techniques

(continued)

Table 7.1 (continued)

Issue	Description
How can decision-makers better link settings, experiences and uses?	In a sense, managers produce opportunities for people to experience certain social-psychological outcomes. These opportunities are composed of attributes, combinations of attributes lead to the notion of setting. Attributes are things like rules, regulations, visitor use density, visitor types, amount and type of modification of the natural environment. Combinations of attributes lead to settings that have certain similarities; such settings can be typologized and classified. However, the link between setting attributes and the social psychological outcomes is anything but clear, definitive and deterministic. Indeed, settings represent opportunities in the sense that they facilitate one type of social-psychological outcome over another, but do not ensure that a particular outcome actually occurs. The actual production of the outcome remains with the visitor. A major challenge however, is to increase our understanding of how settings, experiences (the package of social-psychological outcomes produced by the visitor) and other uses are linked. Increasing our understanding would allow more efficient and mindful allocation of opportunities to settings
What is the role of tourism as a component of a community's economy?	Shifts in economic restructuring have increased not only the economic importance of tourism in protected areas but also have changed relationships between communities and adjacent public lands. As employment and revenue from traditional resource commodity processing has dropped, many communities have turned to tourism as a tool to maintain their economic and social vitality. For many of these communities, however, the product sought by non-resident visitors is located on publicly administered lands. Understanding what economic role tourism may hold in a community is difficult because many businesses in the tourism sector also appeal to residents, such as restaurants, service stations, lodging. Sorting out what is attributable to tourism is difficult
How do changes in the amount, location and character of the human population impact formulation of policy?	Many areas have experienced dramatic population changes over the last 15 years. These changes have generally resulted from significant in-migration, particularly to rural areas and more specifically into areas high in amenity value. Such population growth has brought generally younger, less affluent individuals into these areas, but individuals also with a different distribution of intellectual skills, social preferences, and activism than the current residents. This population growth also accompanies structural shifts in regional economies, generally from manufacturing, natural resource dependent economies to service and amenity dependent economies. In the U.S., such population growth is relatively widespread in counties with a large proportion of land managed by the federal government: About 94 % of the counties in the U.S. with more than 30 % of their land base in federal stewardship saw significant population growth in the 1990s
What public land sustainable tourism opportunities should be commercialized and privatized?	With decreasing budgets and more conservative political philosophies, managers are under more pressure to commercialize tourism opportunities on public lands. Such commercialization and any accompanying privatization would increase the costs to the recreating public, raise expectations of the quality of opportunity to be provided, and increase revenues to management. But which opportunities should be commercialized? What criteria would be used to make this decision?

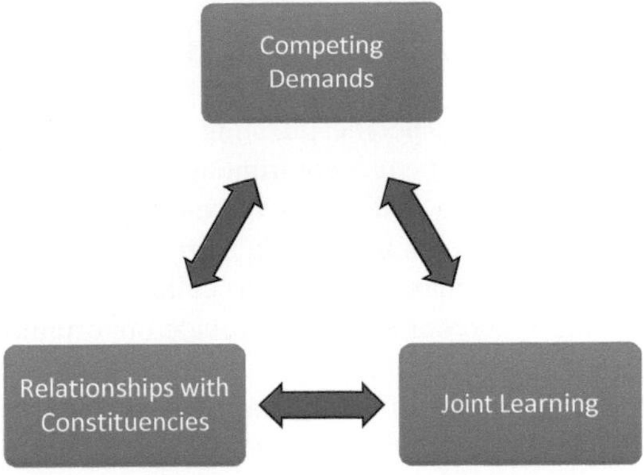

Fig. 7.2 In twenty-first century sustainable tourism, protected area managers deal with three principle task as suggested here. Managing these tasks and their interactions is a major challenge (Suggested by McCool and others 2013)

by forcing us to explicate and "work through" the various dimensions of them. A recreation planning framework helps decision makers gain insight about the particular issue confronting them and then provides some guidance on how to address the issue.

More specifically, framework here is defined as a process that is focused on recreation consisting of a group of guidelines, propositions, or steps that help frame or define the problem, forcing explicit consideration of issues and consequences. We exclude from this definition various laws and regulations that prescribe particular processes or mechanistic formulae that lead to specific answers. Stankey and Clark (1996) suggest that an effective framework would: (1) identify trade-offs between provision of recreation opportunities with the resulting local economic impacts and protection of biodiversity values; (2) appreciate and address complexity (rather than suggest reductionistic approaches), and (3) accommodate the array of constituencies with interests in the specific area or issue.

An example is provided by McCool and others (2013) when they suggest that protected area managers, including those faced with providing opportunities for sustainable tourism, deal with three major tasks: (1) managing competing demands; (2) managing relationships with their constituencies—which often articulate those demands; and (3) managing learning (see Fig. 7.2). This framework help managers understand that these tasks are connected and that focusing on them and their interactions helps to comprehend the system within which sustainable tourism is embedded.

7.4 What Criteria Are Useful in Assessing the Suitability of a Planning Framework?

Not all recreation and tourism planning frameworks are suitable for all issues confronting protected areas. And there may yet be issues for which no suitable framework exists. Never the less, decision-makers must evaluate the suitability of a framework for a specific issue. I suggest five criteria to assess the suitability of a recreation and tourism framework in any given situation.

A primary consideration is the *saliency* of the framework to the particular problem in a specific planning situation. Not all frameworks were designed to address all challenges confronting protected area tourism planners. Indeed, as shown in Table 7.1, a wide range of challenges exist out there. Therefore, as a first step, a framework should provide a process for working through the specific challenge confronting managers. In particular, the framework should help clarify the issue, frame it appropriately and provide opportunities to think critically about how the issue can be resolved.

The next set of criteria are adapted from (Brewer 1973). The framework should be *conceptually sound*, that is it should be based on the most current and appropriate science and theory. Use of the framework should be relatively easy to defend to one's peers. The framework should also meet certain *technical* criteria, that is, it should be easily translated into practice. One should be aware of the key knowledge, skills and abilities required to implement the framework and it should be within the capacity of the organization to implement the framework.

The framework must meet an *ethical* criterion as well, that is, it should identify who wins and who loses, the distributional consequences of a decision. Finally, the framework must be *pragmatic*, that is it must be both efficient (getting the biggest bang for the buck) and it must be effective, (it helps achieve larger goals, such as optimizing the flow of benefits from public lands).

7.5 What Frameworks Are Found in the Planner's Toolbox?

A limited number of frameworks exist to assist protected area tourism managers to address twenty-first century issues. These are listed in Table 7.2. The frameworks represent an evolution of not only how sustainable tourism issues are addressed but also in how they are framed. The purpose here is not to present each framework or evaluate them in depth (see (McCool and others 2007) for a comparative assessment), but rather to depict the issues associated with use of recreation planning frameworks.

Recreation planning issues have undergone a major evolution in how they are cast as represented by the continual development of new frameworks from the learning engendered by application of frameworks in the past. A primary example is the carrying capacity approach[1] to visitor impact management. Carrying capacity has often been viewed as the fundamental question underlying tourism and visitor management in protected areas, and has been frequently defined as the amount of recreational use that can be sustained without degradation of the biophysical

[1] I hesitate to include carrying capacity as a framework because it meets none of the criteria identified earlier, but it has nevertheless dominated recreation management for decades.

Table 7.2 List of recreation and tourism management frameworks, the principal question each addresses and key references

Framework	Principal question	Key references
Recreation opportunity spectrum based frameworks Recreation opportunity spectrum—1970s Tourism opportunity spectrum—1990s Water recreation opportunity spectrum—2000s	What settings exist and what should be provided?	Clark and Stankey (1979) Driver and Brown (1978) Dawson (2001) Haas et al. (2004)
Limits of acceptable change based frameworks Limits of acceptable change—1980s Visitor impact management—1980s Visitor experience and resource protection—1990s Visitor experience and resource protection—1990s Tourism optimization and management model—1990s	How much change from natural conditions is acceptable?	Stankey et al. (1985) Graefe et al. (1990) Hof and Lime (1997) Manidis (1997)
The benefits based management framework—1990s	What experiences should be provided?	Driver and Bruns (1999)
Carrying (visitor) capacity based frameworks—1960s + Social Social Biophysical Facility	How many is too many?	Lime and Stankey (1971) Haas (2002)
Placed-based frameworks—2000s	What meanings are attached to this place?	Kruger and Jakes (2003)

attributes of the area or the experience constructed by visitors (see McCool and Lime 2001 for a critique of recreational carrying capacity). Driven by a desire to manage impacts, managers have often sought the answer to the question "How many is too many?" Unfortunately, answers have been few.

Ultimately, the issue of carrying capacity is one of how the challenges of tourism and visitor management, use and impacts are framed. Too often, planners attempt to solve the wrong problem, solve solutions, or state the problem in such a way that it cannot be solved (Bardwell 1991). Carrying capacity represents this situation. The concern about protected area recreation and tourism is a set of doubts dealing with the amount and kind of impacts that are generated, the ability of a site or community to assimilate impacts, the acceptability of impacts—both social and biophysical—, the trade-offs made under conditions of uncertainty, the ability of those affected to participate in decisions, the institutional capacity to monitor and manage impacts over time, and the will of the political system to make often difficult and controversial decisions. This complex assortment of interacting issues cannot be successfully reduced to the question of "How many is too many?"

Our failures in finding a carrying capacity have taught us by reframing this question, we more closely get at the intention reflected in it: What are the desirable, appropriate or acceptable conditions for this region, area or tourism destination? Once that is decided, we can then discuss how different management practices meet the tests of efficiency, effectiveness and efficacy (Checkland and Scholes 1990) that are important criteria in evaluating resolutions to messy problems.

Table 7.2 also shows the principal question addressed by each of the frameworks. Note how the frameworks differ significantly on this point, suggesting that each framework serves protected area tourism managers in different ways and that each framework varies in its suitability in addressing current and anticipated issues.

7.6 Conditions Needed to Implement a Recreation Planning Framework

Of course the frameworks listed in Table 7.2 can only be implemented if a set of conditions are present in the agency considering using a framework. These are briefly discussed below.

The agency must have the *organizational will* to implement the framework in full. The frameworks listed in Table 7.2 consist of a sequence of steps, elements or components. Each of these is essential to successful completion of the framework, and thus resolution of the underlying problem. Often, I have been asked how can the framework be shortcutted, that is steps dropped, indicators and standards borrowed from other areas and so on. While this might seem a good way to cut costs in the short run, planning frameworks are not about the short run—they are about learning, thinking about the future, engaging the public, and strategic analysis. Each step or element is included for a specific reason and reflects an underlying principle; dropping any out is counter to the notion of planning as the application of critical thinking. Thus, the organization most importantly must have the determination to complete the process.

Related to this condition is that the personnel involved cannot be rushed, careless, or distracted. They must have the time and resources to complete a planning process competently. This would require the organization to develop and make available the time needed to work through the challenges of a sustainable tourism issue. Careful attention to personal values toward the environment as well as the role of communities in decision making are also essential.

Second, the organization needs the *technical capacity* to conduct the planning processes. By this, I mean the organization needs the personnel with the appropriate skills, some technical, some in public meeting facilitation, and many other conceptual skills. This means the organization must not only seek out trained individuals, but also engage in in-service training and continuing education to maintain an up-to-date work force.

Third, the process must be *inclusive* of differing values and systems of knowledge. Many decisions in tourism management are value judgments (Krumpe and McCool 1998) and thus a full discussion of the values involved is essential to addressing sustainable tourism problems. This can only be done with inclusive public engagement processes because technical planners cannot be expected to equitably represent every value system. In addition, there must be recognition that different forms of knowledge (e.g., experiential, scientific) are not only legitimate ways of knowing but each contributes constructively at varying points in a planning process.

Fourth, the process must be *open and deliberative*, with opportunities to express, challenge, and debate varying assumptions underlying proposed actions and goals. This characteristic is essential not only to learning but to building trust and ownership in the protected area and the planning process. The planning process must therefore secure safe and accessible venues, ones that symbolize equality of access and so on.

Fifth, the process should focus on *effectiveness* of the framework not just efficiency. Here, I mean that attempting to keep costs low should be viewed within the context of what needs to be done. Often, for example, public engagement is viewed as an "added cost" to protected area planning, with meetings are often perceived as simply a means of collecting data about public preferences rather than also opportunities for building networks, strengthening relationships, constructing consensus and fashioning trust among constituencies.

Finally, these frameworks will work most effectively when we think at the *systems level*. Systems thinking involves considering relationships across time, space and function. By using systems thinking, we identify leverage points, temporal delays, constituencies and second and third order consequences of actions. Such thinking recognizes the complexity of the sustainable tourism world and we benefit of employing this thinking in developing more effective responses to challenges and opportunities as well as more efficiency in our management (McCool and others 2015).

7.7 Conclusion

Managers of protected areas, including those charged with administering sustainable tourism opportunities operate in a complex, dynamic and messy environment, where competing goals and a lack of science challenge their ability to frame problems and develop responses. Sustainable tourism management is as much about the values and preferences of constituencies—including community members, visitors, tour operators, environmentalists, government officials—as it is about technical concerns of biophysical impacts. Frameworks help managers work through these problems by structuring thinking processes, explicating assumptions and values, and forcing consideration of a range of consequences, interests, and alternatives. Frameworks help managers gain useful insight and frame an issue or problem in productive ways. The implementation of a framework in any specific situation is predicated on the presence of a number of organizational, personal and technical conditions. Of primary importance is the organizational will to implement and competently complete a particular framework.

If a framework can be viewed as an innovation, then adoption of this innovation follows a certain, and generally predictable, path (Rogers 1995). Rogers argues that the adoption of an innovation by a member (say a manager) of a social system depends heavily on the decisions of other members of the social (managerial) system. We expand this to include the *experience* of other members of the system with the particular innovation. The close collaboration of scientists and managers that typified the development of the ROS and LAC frameworks allowed managers to adopt the innovation in small steps, and with the support of scientists.

Positive experiences of other members of the social system (in Rogers' language, innovators and early adopters) provide the confirmation that the innovation will enhance a person's ability to function effectively. Such experiences reduce the risk of adopting an innovation and then it failing. This process certainly occurred with the Limits of Acceptable Change system following its use in the Bob Marshall Wilderness in the mid-1980s. Other managers, once they heard of the use of LAC there, frequently called and asked for information, why the process was successful and for help in adopting it for their own areas.

The Recreation Opportunity Spectrum framework was successful because it helped managers understand and integrate recreation into decisions in a multiple use situation. LAC succeeded because it helped managers structure their thinking about the trade-offs between partially conflicting goals.

However, innovations and bureaucracies are polar opposites. Bureaucracies, such as protected area and tourism management agencies, are established to deal with routine problems and issues. Protected area agencies are notoriously conservative, with a top-down command and control structure. In these situations, innovations are anything but routine and are slow to come. Diffusion strategies must emphasize, Rogers argues, the compatibility of the innovation with existing agency norms and policies.

While this paper has described the frameworks of the past, the paper itself is ultimately about the future, about the issues and challenges the new century poses and the capability of protected area agencies to respond to them and provide high quality opportunities for sustainable tourism. There is no question that protected area organizations are experiencing an era of declining capacity. This decline is measured in terms of both the managerial proficiency (available to sustain tourism and visitation opportunities) and the scientific expertise needed to support the information requirements good stewardship requires. The loss of both types of capacity has long term consequences to the ability of an organization to respond to evolving issues and challenges.

Successful framework applications have occurred as a result of close, continuing collaboration between managers and scientists. Such collaboration allows managers to communicate issues and mandates clearer to scientists, scientists can query managers and come to a better understanding of the job at hand, and as a result develop applications, concepts and processes that are more useful to managers.

Approaches to recreation and tourism development issues that have not involved this collaboration, in general, have not had widespread application. Given declines in both types of capacity, one can only wonder if public land recreation management agencies contain the capacity to address the issues of the new century. Undoubtedly, if the response is "no", the public eventually will demand changes in agency budgets, priorities and mandates.

Given the complex, contentious and changing environment in which tourism development decisions are being made, there is a need to continually understand the strengths and weaknesses of these frameworks, to monitor the situations in which they work or don't work, and to periodically make changes in how they are implemented. Frameworks, such as those briefly mentioned here, can help managers dive deeper and think differently about sustainable tourism, about how they can see sustainable tourism achieve long term goals of protecting heritage, advancing quality of life and building community resilience.

The limited capacity of protected area agencies to manage tourism and visitation inevitably leads to the conclusion that decision-makers must have a better understanding of these frameworks, the key concepts and assumptions upon which they are built, and their suitability for addressing different issues. Developing a capability in these areas ultimately will lead to more efficient, effective and equitable decisions.

While a number of frameworks do exist, not all of them necessarily address the issues confronting sustainable tourism management in the twenty-first century. Frameworks have improved the quality of management in addressing many issues in the past. The issues with no frameworks are crying for development of new ones.

References

Bardwell, L. 1991. Problem framing: A perspective on environmental problem-solving. *Environmental Management* 15(5): 603–612.

Borrie, W.T., W.A. Freimund, and M.A. Davenport. 2002. Winter visitors to Yellowstone National Park, their value orientations and support for management actions. *Human Ecology Review* 9(2): 41–48.

Brewer, G.D. 1973. *Politicians, bureaucrats, and the consultant: A critique of urban problem solving*. New York: Basic Books.

Checkland, P., and J. Scholes. 1990. *Soft systems methodology in action*. West Sussex: Wiley.

Clark, R.N., and G.H. Stankey. 1979. *The recreation opportunity spectrum: A framework for planning, management and research*, PNW-98. Portland: USDA Forest Service, Pacific Northwest Forest and Range Experiment Station.

Dawson, C.P. 2001. Ecotourism and nature-based tourism: One end of the tourism opportunity spectrum? In *Tourism, recreation and sustainability: Linking culture and the environment*, ed. S.F. McCool and R.N. Moisey, 41–53. Wallingford: CABI.

Driver, B.L., and P.J. Brown. 1978. The opportunity spectrum concept and behavior information in outdoor recreation resource supply inventories: A rationale. In *Integrated inventories of renewable natural resources: Proceedings of the workshop*, General technical report RM-55. tech. coords. Gyde H. Lund, Vernon J. LaBau, Peter F. Folliott, and David W. Robinson, 24–31. Fort Collins: USDA Forest Service, Rocky Mountain Forest and Range Experiment Station.

Driver, B.L., and D.H. Bruns. 1999. Concepts and uses of the benefits approach to leisure. In *Leisure studies: Prospects for the twenty-first century*, ed. E.L. Jackson and T.L. Burton, 349–369. State College: Venture Publishing.

Dustin, D.L., and I. Schneider. 2005. The science of politics/the politics of science: Examining the snowmobile controversy in Yellowstone National Park. *Environmental Management* 34(6): 761–767.

Graefe, A.R., F.R. Kuss, and J.J. Vaske. 1990. *Visitor impact management: A planning framework*. Washington, DC: National Parks and Conservation Association.

Haas, G.E. 2002. *Visitor capacity on public lands and waters: Making better decisions. A report of the federal interagency task force on visitor capacity on public lands*. Submitted to the Assistant Secretary for Fish and Wildlife and Parks, U.S. Department of the Interior, Washington, DC, 1 May 2002. Published by the National Parks and Recreation Association, Ashburn, 42 p.

Haas, G., R. Aukerman, V. Lovejoy, and D. Welch. 2004. *Water recreation opportunity spectrum (WROS) users' guidebook*. Denver: U.S. Department of the Interior, Bureau of Reclamation.

Hof, M., and D.W. Lime. 1997. Visitor experience and resource protection framework in the national park system: Rationale, current status, and future direction. In *Limits of acceptable change and related planning processes: Progress and future directions*, ed. S.F. McCool and D.N. Cole, 29–36. General technical report INT371. Ogden: USDA Forest Service, Intermountain Research Station.

Kohl, J., and S.F. McCool. 2016. *The future has other plans*. Denver: Fulcrum.

Kruger, L.E., and P.J. Jakes. 2003. The importance of place: Advances in science and application. *Forest Science* 49(6): 819–821.

Krumpe, E.E., and S.F. McCool. 1998. Role of public involvement in the Limits of Acceptable Change wilderness planning system. In *Limits of acceptable change and related planning processes: Progress and future directions*, ed. S.F. McCool and D.N. Cole, 16–20. INT-GTR-371. Missoula: USDA Forest Service Intermountain Research Station.

Lime, D.W., and H. Stankey. 1971. Carrying capacity: Maintaining outdoor recreation quality. In *Recreation symposium proceedings*, 174–184. Upper Darby: Northeastern Forest Experiment Station, USDA Forest Service.

Manidis R. Consultants. 1997. *Developing a tourism optimization management model (TOMM), a model to monitor and manage tourism on Kangaroo Island, South Australia*. Surry Hills: Manidis Roberts Consultants.

Manning, R.E. 2004. Recreation planning frameworks. In *Society and natural resources: A summary of knowledge*, ed. M.J. Manfredo, J.J. Vaske, B.L. Bruyere, and P.J. Brown, 83–96. Jefferson: Modern Litho.

McCool, S.F., and D.W. Lime. 2001. Tourism carrying capacity: Tempting fantasy or useful reality? *Journal of Sustainable Tourism* 9(5): 372–388.

McCool, S.F., R.N. Clark, and G.H. Stankey. 2007. *An assessment of frameworks useful for public land recreation planning*. General technical report PNW-GTR-705. Portland: U.S. Department of Agriculture, Forest Service, Pacific Northwest Research Station, 125 p.

McCool, S.F., Y. Hsu, S.B. Rochas, A.D. Sæþórsdóttir, L. Gardner, and W. Freimund. 2012. Building the capability to manage tourism in support of the Aichi targets. *Parks* 18(2): 92–106.

McCool, S.F., A.B. Nkahta, C. Breen, and W.A. Freimund. 2013. A heuristic framework for reflecting on protected areas and their stewardship in the 21st century. *Journal of Outdoor Recreation and Tourism* 1: 9–17.

McCool, S.F., W.A. Freimund, and C. Breen. 2015. Benefiting from complexity thinking. In *Protected area governance and management*, ed. G.L. Worboys, M. Lockwood, A. Kothari, S. Feary, and I. Pulsford, 291–326. Canberra: ANU Press.

Moore, S.A., A.J. Smith, and D.N. Newsome. 2003. Environmental performance reporting for natural area tourism: Contributions by visitor impact management frameworks and their indicators. *Journal of Sustainable Tourism* 11(4): 348–375.

Nilsen, P., and T. Grant. 1998. A comparative analysis of protected area planning and management frameworks. In *Limits of acceptable change and related planning processes: Progress and future directions*, ed. S.F. McCool and D.N. Cole, 49–57. Ogden: USDA Forest Service Rocky Mountain Research Station.

Nkhata, A.B., and S.F. McCool. 2012. Coupling protected area governance and management through planning. *Journal of Environmental Policy and Planning* 14: 394–410.

Rogers, E.M. 1995. *Diffusion of innovations*. New York: Free Press.

Stankey, G.H., and R.N. Clark. 1996. Frameworks for decision making in management. In *Proceedings of the 1996 World Congress on Coastal and Marine Tourism*, ed. M. Miller, 55–59. Honolulu: University of Washington.

Stankey, G.H., D.N. Cole, R.C. Peterson, M.E. Lucas, and S.S. Frissell. 1985. *The limits of acceptable change (LAC) system for wilderness planning*. General technical report INT-176. Ogden: USDA Forest Service.

Weick, K.E., and K.M. Sutcliff. 2001. *Managing the unexpected: Assuring high performance in an age of complexity*. San Francisco: Jossey-Bass.

Part III
Case Studies

Chapter 8
When 'dem Come: The Political Ecology of Sustainable Tourism in Cockpit Country, Jamaica

Jason A. Douglas

Abstract Sustainable livelihood projects in the Caribbean have been largely unsuccessful to date, often culminating in a failure to fulfill initial conservation and development objectives. This chapter examines the political ecology of sustainable tourism in the bauxite rich Cockpit Country of West Central Jamaica. It concerns the collaboration of government and non-government organizations and people living in forest-fringe communities throughout Cockpit Country working toward sustainable alternatives to bauxite prospecting and mining that stakeholders anticipated would come to fruition in the area. This chapter focuses on the operations of, and concerns raised by, various alternatives to bauxite mining and the implementation of these alternatives throughout Cockpit Country.

Keywords Caribbean • Jamaica • Cockpit country • Sustainable tourism • Political ecology

8.1 Introduction

Jamaica has been deeply inserted in global trades and processes, from sugar cane production to bauxite (the raw material used in aluminum production) mining, since it was colonized in the seventeenth century. A recent and localized testament to this concerns the Cockpit Country of West Central Jamaica (Fig. 8.1), which has increasingly become a space and place embedded in discourses of conservation, development, and poverty alleviation over the course of the last decade. It is a place that has attracted research concerning the rich flora and fauna that are endemic to Cockpit Country, as-well-as a space where local residents form their livelihoods, predominantly through agricultural practices. Simultaneously, Cockpit Country is part of an area, which includes Florida and the Caribbean, that has been described as the world's third most important biodiversity "hotspot" (TNC 2007).

J.A. Douglas (✉)
Department of Environmental Studies, San Jose State University, San Jose, CA, USA
e-mail: jason.douglas@sjsu.edu

S.F. McCool, K. Bosak (eds.), *Reframing Sustainable Tourism*,
Environmental Challenges and Solutions 2, DOI 10.1007/978-94-017-7209-9_8

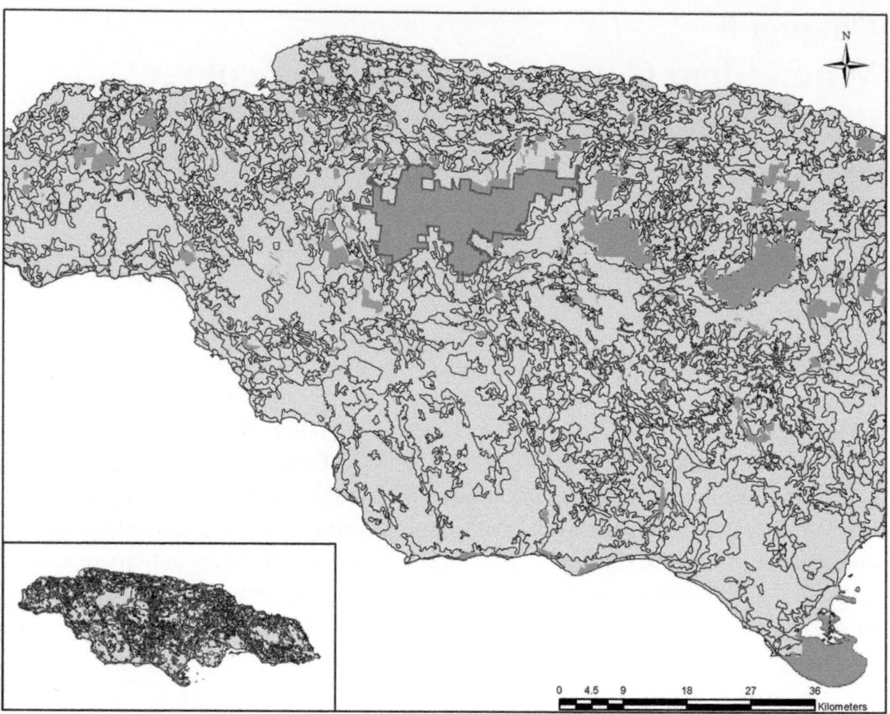

Fig. 8.1 Cockpit country, Jamaica

This chapter seeks to address the insertion of Cockpit Country in such global processes in the context of bauxite mining and sustainable tourism, whereby the natural resources, biodiversity, and culture of the area present highly sought after market commodities. The current case study concerns the establishment of sustainable tourism and niche market areas in Cockpit Country as sustainable alternatives to bauxite prospecting and the potential of ensuing mining by the Aluminum Company of America (ALCOA) and the Jamaican Almuminum Company (JAMALCO). Key to an analysis of the insertion of Cockpit Country in global processes is the development of a relational understanding of the production of space, place and power as situated in sustainable tourism involving private lands and public projects. To build on this analysis and add to the broader political ecology literature, I argue that it is essential to analyze the nuanced ways in which people produce their understandings of nature and society in the context of sustainable tourism and sustainable development more broadly. To begin, I will present a critique of people's insertion into the sustainable development discourse and practice in an effort to frame an understanding of the broader problems and potentials of sustainable tourism. Building on this critique, I will present a case study concerning the development of a participatory forest conservation program—which involved agroforestry, reforestation, and ecotourism—as situated in a broader conservation and development program involving Local Forestry Management Committees (LFMC) in an effort to develop alternatives to bauxite mining and traditional agricultural practices.

8.2 People and the Sustainable Tourism Network

People living in areas immersed in environmental conflict, from mineral mining to broader land use rights, such as the Cockpit Country of Jamaica, have increasingly expressed concern over the neoliberal practices of northern institutions. NGOs and academic research institutions have been instrumental in developing alternatives to neoliberal practices that are inherent to the dominant development discourse and practice; however, such institutions and practices have also been identified as proxies for the extension of neoliberalism, whereby they present compromises between the neoliberalism of the global north and the development of social movements in the global south. NGOs, for example, seek compromises that in-effect only further advance neoliberal policies; that is, their ideology and practice can be quite distant from practices that would redress sources of poverty and offer solutions to environmental inequities. Further, such organizations tend to be organized in a top-down manner, whereby their solutions ultimately marginalize the voice and agency of the people and communities they work with (Brosius et al. 1998; Douglas 2013; West 2006). Through the ensuing analysis, I argue that such marginalization is clearly demonstrated in people's nuanced experiences of nature and society; that is, people's material and conceptual production of nature (Douglas 2014; Smith 2008). For example, NGOs tend to promote market and service based solutions, e.g., ecotourism, instead of developing transparent, bottom-up strategies to manage environmental resources based on local people's experiential knowledge. This approach suggests that the environmental and broader social issues that people in the global south are faced with revolve around individual practices rather than broad-scale institutional practices predominantly stemming from northern institutions. Meanwhile, the powerful actors in such systems maintain power, while appropriating the labor and invaluable local environmental knowledge of rural populations, whose work and intricate knowledge are not only exploited but are appropriated in the commodification of nature (Hale 2002).

The people participating in the LFMCs come from various social standings, from people who live and work in urban Jamaica and abroad to the rural communities of Cockpit Country. It is this latter set that the LFMCs are predominantly concerned with, environmentally and socially. Local farmers, Maroons, and a host of people who practice various trades in Cockpit Country—people who are considered to be in the "lower sets" (Thomas 2004) of Jamaican society—are situated in a paradigm of race, class, and gender that reflects the conditioning of local circumstances by actions occurring at other scales of the global economy, e.g., economic restructuring and the downturn in the global economy. They are immersed in a neoliberal capitalist development program that unequally structures access to resources and opportunities (Thomas 2004). Jamaica, like so many southern nations, underwent IMF structural adjustment programs in the 1980s and continues free trade small state policies to current times (Polanyi-Levitt 1991; Robotham 2005; Thomas 2004; Weis 2000), where the majority of profits from nature are directly deposited to northern institutions, leaving little benefit for the people of Jamaica (Polanyi-Levitt 1991). Further, the increase of cheap food imports over the last 30 years has effectively disenfranchised small-scale farmers from domestic markets (Weis 2000).

In the Jamaican context, unequal access to social and environmental resources is fundamentally rooted in the structures of race, class, and gender that took shape in the British colonial plantation system. These colonial conditions have been further exacerbated by the downturn in the global economy, whereby the exploitation of the Jamaican working class people and environment is being repositioned in the sustainable development paradigm, in many ways serving to commodify nature (Douglas 2013). Yet, it is particularly important to note that the inhabitants of Cockpit Country have a practical relationship with the environment through farming, the primary form of employment there (McGregor et al. 1998), and direct interaction with the local environment more broadly. Practical engagement is likely to be important in shaping people's understandings of their surroundings (Carrier 2003; Kellert et al. 2000). Programs like the Cockpit Country LFMC natural resource participatory management program link "local populations, national agencies, and international organizations" (Orlove and Brush 1996). These linkages bring together people with very different material and conceptual understandings of nature. Considering this, we must begin to ask how such linkages materialize in the context of integrating conservation with the development of ecotourism programs and the like. This chapter provides and ethnographic perspective of the problems and potentials of the processes of international and local NGOs, international aid agencies, national government agencies, and local communities working together in the development of sustainable tourism projects intended to promote forest resource conservation and poverty alleviation. Furthermore, I will elucidate people's nuanced understandings of nature and society as a fundamental unit of analysis in unpacking the problems and potentials of sustainable tourism in Cockpit Country.

8.3 The Research Context

The 1996 Forestry Act of Jamaica (Forestry Department 1996) was introduced in an effort to harness and promote the tenets of sustainable development that were laid out in Agenda 21, a product of the 1991 Rio Earth Summit. While Jamaica had to face a stark reality concerning deforestation stemming from the colonial era (Evelyn and Camirand 2003; Wimbush 1935), which ravished the countries natural resources, it also acknowledged that the country was struggling to maintain its place in the global economy, particularly after the collapse of the sugar industry in the 1980s (Polanyi-Levitt 1991). Therefore, dedicating funding toward the ends of conservation was more of a luxury than a necessity (Lundy 1999). Furthermore, with the recent collapse of the bauxite industry, also an environmentally and socially destructive practice, there has been increasing motivation on the part of the Jamaican government to work with allying agencies intent on promoting sustainable industries that may continue to circulate Jamaica's natural resources in the global economy. As a result, the Forestry Department of the Ministry of Agriculture (FDJ) embraced the language of sustainable development in an effort to sustain the countries natural resources in economically and environmentally sustainable ways. This, as proposed in the 1996 Forestry Act, led to the inclusion of the Jamaican populace in the conservation discourse, leading to the establishment of the Cockpit Country Local Forestry Management Committees (CCLFMCs).

8.3.1 Cockpit Country Local Forestry Management Committees

Beginning in 2007, Cockpit Country Local Forestry Management Committees were formed in response to prospecting practices supported by the Ministry of Energy and Mining and the Jamaican Bauxite Institute. A Special Exclusive Prospecting License (SEPL 535) and an Exclusive Prospecting Licenses (EPL 536) were granted by the Ministry of Energy and Mining for various areas of Cockpit Country beginning in 2004. In response to the granting of these licenses, a host of environmental organizations allied to form the Cockpit Country Stakeholders Group (CCSG) in an effort to educate the national and international community on the effects of mining and curb bauxite prospecting practices in Cockpit Country, which the CCSG argued to be a threat to the nature and communities of Cockpit Country (Dixon 2006). These organizations realized that they would garner little interest from government offices without the support of the people of Cockpit Country. To address this issue, they began to intensify efforts in environmental education and community outreach. With a well-established interest in sustainable forest resource use in Cockpit Country and Jamaica more broadly, the FDJ also claimed a stake in working toward community-based solutions.

With funding from USAIDs Parks in Peril (PiP) project (TNC 2007), The Nature Conservancy began working with the CCSG and the FDJ to seek out community interest in, and ideas for, establishing a series of Local Forestry Management Committees (LFMC) throughout Cockpit Country. Their primary purpose was to establish groups that would assist in forest conservation. A similar initiative had been attempted by the World Bank in 1999, but they withdrew the initiative due to community and government resistance to their proposal, which effectively would have limited all community activity in the forest alongside banning bauxite mining (World Bank 1999). Throughout 2007, the CCSG held a series community fora in various areas of Cockpit Country including Albert Town, Trelawny; Bunkers Hill, Trelawny; and Flagstaff, St. James, as well as in more high profile areas including resort hotels in Montego Bay and Kingston. While the communities in these areas greeted these organizations with caution at first, placing emphasis on the possible benefits for the community members, they were eventually drawn into the LFMC organizational architecture, whereby the notion of a democratic and participatory process concerning the development of conservation and alternative livelihood programs was eventually embraced. The proposed collaboration was rooted in the commonly accepted belief that bauxite mining would be detrimental to the people and land of Cockpit Country. Some of the additional risks identified included unsustainable farming practices, endangered species collection, and improper sanitation (TNC 2006).

Three LFMCs were officially launched in 2008 around the Cockpit Country periphery. The LFMCs were strategically set up in these locations so that they could effectively reach out to the surrounding communities and form a social buffer zone, as the Government of Jamaica (GoJ) had not responded to CCSG requests to establish a buffer zone for the purposes of delimiting bauxite mining areas. With continued funding garnered from USAID's Protected Areas and Rural Enterprises (PARE) project (USAID 2009), the newly established LFMCs provided training in the areas of small business development, ecotourism, food preparation, and cultural sensitivity for their Cockpit Country based participants. A significant portion of this funding—$12 million J ($118,500 US)—was allocated to the development of an ecotourism visitor center in Flagstaff. The ecotourism project development included the rehabilitation of Maroon historical sites and a series of trails that had become overgrown due to infrequent use in Flagstaff, St. James. As the PARE funding ended in October of 2009, TNC and USAID ceased direct involvement in the project and handed all responsibility over to the FDJ and the community members that had helped to establish the CCLFMC project.

Since all projects that began with the CCLFMCs, FDJ, TNC, and USAID were entirely left to the efforts of the CCLFMC members with the assistance of the FDJ, it was hoped that the training provided would serve as the basis for the LFMC members to begin to garner their own funding sources and continue to forge ahead in the completion of their initial conservation and alternative livelihood projects, namely ecotourism. It is the product of 3 years of direct work in Cockpit Country through the collaboration of the LFMCs with USAID, TNC, FDJ, and other NGOs and research organizations that concerns this chapter.

8.4 Working Together for "Sustainable" Tourism

> To save nature all of us need to work together: development agencies, governments, private sector partners, and—most importantly—the local communities whose livelihoods are at stake (USAID 2010).

The language of USAID is one grounded in concepts of participatory forest conservation, whereby powerful national and international organizations work with disenfranchised forest fringe communities to sustain the resources in the areas where they live, work, and play. This is a language of community partnerships created to promote a space, place, and practice that appeals to a broad range of people and their concerns for nature. The quote in the head of this section proposes strategic partnerships between groups of people that are situated in very different social and environmental circumstances.

To illustrate these processes, a series of events stemming from the USAID and TNC flagship project based in Flagstaff, St. James will be described. In 2009, TNC redirected all funding for ecotourism activities to the Flagstaff LFMC (SWLFMC). This sudden and unexpected shift left the Bunkers Hill LFMC (NLFMC) with no funding or development initiatives. Furthermore, the Albert Town LFMC (SELFMC) participants were concerned that their group did not receive the same opportunity as Flagstaff, particularly since the Bunkers Hill LFMC had lost all of their funding. These events stood in tension with the LFMC participants understanding of the LFMC process—the LFMCs were established as democratic organizations, whereby all participants shared a "stake" in decision-making processes regarding project design, implementation, and funding allocation. However, it was clear the high level funding decisions were in the power domain of the agency that administered training and allocated USAID funding as such, allowing TNC to dictate LFMC practices. This was a fundamental breach of the understanding negotiated among the respective agencies and community members held at the initial meetings.

Given this breach of the LFMC process, community members were not afforded "voice" concerning the development of ecotourism and the like in Cockpit Country. While establishing the LFMCs and their affiliated programs, the language of sustainability was employed to create an environment of "stakeholder partnerships" where the agencies that were working with community members in the LFMCs maintained a firm grasp on the "stake" that LFMC members had been promised. This was illustrated by the fact that LFMC participants had been assured that they would participate in all decision-making and that profits from ecotourism ventures would directly benefit the communities. However, TNC did not engage in the LFMCs democratic process to determine where the benefits of ecotourism should land. In this case, the greatest benefits were awarded to TNC for administering the project—a benefit that the LFMC participants had no knowledge of. For the community of Bunkers Hill, the establishment of a sustainable tourism enterprise in Flagstaff raised feelings of resentment toward the Flagstaff community and a landowner and LFMC participant in the NLFMC, who refused to have his land exploited for the benefit of a TNC project. Echoing clientelistic relationships common in the area, community resentment was redirected toward the landowner in the NLFMC area and the SWLFMC that received the ecotourism funding, rather than TNC.

One of the predominant issues in this case concerned access to land and resources. When speaking with community members about the possibility of developing ecotours in the Cockpit Country Forest Reserve, there was often an expression of the tension between private property and public projects. For example, one person speaking on reforestation said:

> 'Dem [FDJ] have no right fi go down there still. You haffi walk 'pon people land fi go in de forest. That ca'an work! Forestry a one road go in. You ca'an drive go in a forest. You haffi walk 'pon, if you walk here so [pointing to a forest access point] fi go down a forest, you

haffi walk 'pon people land fi go a inna forest, you haffi go through a person land fi go inna forest . . . . Yeah, that a one big problem.

This tension between private land and public projects was quite clear in the sentiments of the Bunkers Hill community. The following interview excerpt from a landowner and LFMC participant in Bunkers Hill illustrates some of the issues raised by land agreements and disputes:

. . . I said to them [TNC], "look, if you are going to use my property, then it would be fair for me to benefit, because it wouldn't be good because 10 years, 15 years we see people getting fat from it and I am still there." If we see, and it also came back I'm saying it doesn't make sense for you to use my property at the end of the day, the benefit I have got is minimal compared to what other people have got. No, it can't be fair; it must be equitable! So these were some of the issues that I had to deal with.

The NLFMC participant inevitably ceased all involvement with the LFMC due to community tension compounded by his disagreement with TNC officials concerning the use of private property in Bunkers Hill. At this point, he had lost his ability to act as a conduit for negotiating the development initiatives with USAID and TNC, which had promised to bring much needed investment to the community. However, this relocated opportunity for introducing sustainable tourism to the community was understood by the executive committee members as an inequitable and undemocratic decision on the part of TNC. Due to the failure of the NLFMC business unit in securing a lease that was amenable to TNC terms, the NLFMC members and broader community pegged the landowner as a harbinger of disinvestment in the area. This phenomenon produced a space and place of mystification; one where the local populace was disconnected from the structural inequities—social and environmental—that determined the outcomes of development in their area. Given such contradictions that tend to be inherent in sustainable tourism initiatives, a more situated analysis of this case in comparison with the success of the lease agreement in Flagstaff will be instrumental.

To begin, the private property of an individual in Bunkers Hill was identified as the premier location for ecotourism out of all three LFMC locations. Following an assessment of both Bunkers Hill and Flagstaff by the U.S. Forestry Department in coordination with TNC and USAID, it was determined that Bunkers Hill ranked higher than Flagstaff on a "tourism attractiveness index" with a total ranking of 0.73 (1 being the maximum ranking), which according to USAID, "is remarkable." Given this quantitative realization of Bunkers Hill's attractiveness to the international tourists who desire to consume nature in the global south for the purposes of adventure and the like, a lease was proposed. However, a proposal of 10,000 J (~$105 US) for a 10-year lease was rather unattractive to the landowner. Receiving just 1000 J per year ($11.63 US), the owner exclaimed that he would see "little benefit." The landowner saw his position as being expendable, and he was fully aware of the implications of having a legally binding, long term lease set in place alongside his uncertain position within the LFMC. As the lease agreement was proposed, all proceeds would be deposited into the LFMC account prior to expenditures. The following comment from the landowner provides an important narrative concerning the vulnerability of his position.

> They say, 'oh you're gonna hurt the plants' [in a sarcastic voice], but all of that is crap. You cannot be so overtly conscious about one thing and not overtly conscious about [another] . . . Because most of these environmentalists, they are being fed by these donor agencies giving them money. So they don't have to worry about their food on the table, they don't have to worry about where they are going to get the next meal from.

This situation was not an unfamiliar one to the people of Bunkers Hill and surrounding areas. Speaking with another community member who had some experience with such matters, I was told of an environmental organization that had suggested her property would be an ideal ecotourism area. Based on the program officer's suggestions, the landowner made several improvements to the property, installing a bar and restaurant along with tables, chairs and benches. As it turned out, the program officer, without a formal agreement with the landowner, began to bring groups of 6–10 tourists at a time, charging the tourists 800 J ($8.90 US) per person for the tour. At the end of the day, the tour operator gave the landowner 300 J ($3.30 US) for the use of the property. After she explained this, she exclaimed, "him thief me! Him nah do nuttin' here!" The landowner felt that she and her property were being exploited for the individual gain of a person that fit into the brown middle class of Jamaica. Another family member came out to join the conversation. With respect to working with the LFMCs and using the property for an ecotourism site, the young man said, "mi nuh have a vision of it." He was wary of entering into this type of relationship, where the family was concerned that their property rights and land use may be put into question by the LFMC or related organizations. Instances such as these reflect the deep-seated suspicion of outside interests among the largely landless population of small-scale farmers when it comes to their property rights, or lack thereof. These concerns have their roots in the colonial period (Douglas 2013).

In contrast, the Flagstaff lease was much easier to negotiate. At the same rate of 10,000 Jamaican Dollars over a 10 year period, the lease legally bound the landholders property for the construction of an ecotourism visitor centre on top of a structure that housed a rum bar and the town post office. The structure was located in the town square, where the ecotours would meet before heading out to existing trails on Crown Lands (lands owned by the state). In addition, the landowner agreed to allow the LFMC to develop a small, 10×10 m medicinal plant plot on a piece of his property next to the ecotour trailhead. In contrast to the Bunkers Hill case, the building would not be the center of the tour and would not require any effort on the part of the landowner as far as site maintenance. In fact, the building rooftop provided little benefit to the landowner aside from protection from sun and rain. Also, there were several benefits to be realized through the potential of increased patronage of the rum bar, not to mention a renovation subsidized by USAID funding. At the time, the small piece of land that he allowed the LFMC to use for the plant plot was not in use, and he would benefit from their maintenance of the property.

8.5 Space, Place, and Power

These arrangements call to mind Escobar's (2008) contention that sustainable development mirrors the power structures and top-down processes inherent in the dominant development paradigm, in that they were ultimately negotiated among local landowners and the urban elite. However, while there is a clear unevenness in this case—the groups that were willing to adhere to TNC directives were the winners of development—the LFMC democratic process, a process rooted in the discourse of sustainable development, presents a clear example of Escobar's thesis, in that this process of reconfiguration of the vernacular landscape was negotiated among the urban elite and the local landholders (cf. Massey 2005), yielding very different development outcomes in these geographically distinct locations.

Yet, this singular land lease was contrasted by other attempts to lease lands from other small-scale farmers in Flagstaff. For example, a forestry officer had suggested that a medicinal plant plot should be set up for display at the ecotourism trailhead. While this was initially done on the rum bar owner's land, he decided to renege on that part of the deal so that he could plant bananas, a fundamental part of his livelihood. However, the land at the trailhead, which leads to Crown Lands, was partitioned among various community members. As such, the land for a medicinal plot would need to be rented from another one of the community members. Yet, the land that would have been most appropriate for the plot was owned and used by another community member for agricultural production. As LFMC members pressured the owner to lease them a small portion of the space, it was apparent that she did not want to sacrifice a potentially impressive agricultural yield toward the establishment of ecotourism for an organization with whom she had no affiliation. In essence, the owner might have been stripped of her ability to determine the use of the land in the interests of cultivating medicinal plants that she would not have access to. Considering this case, it is important to note that small landholders associate property ownership with freedom and sovereignty (Weis 2006), and are suspicious of any incursions on that autonomy.

These clear power relations concerning the establishment of ecotourism in Cockpit Country, while thoroughly uneven, recall Cindi Katz's notion of "nature as an accumulation strategy" (1998). One way this process works is to claim the impending demise of nature and note its deleterious effects on global and local populations in order to appropriate these claims in the interests of various potentially profitable preservation strategies such as ecotourism, among others. As Katz rather insightfully noted, the establishment of nature parks, ecotourism ventures, buffer zones and debt-for-nature swaps, all of which materialized to some extent in the Cockpit Country context, follow directly as a solution to the extraction of nature. The wealth and resources typically flow from the nations of the global south to those in the north. In the context of the CCLFMCs, the initiative was to privatize nature through the establishment of public/private partnerships that concentrate capital on spaces marked for agroforestry, ecotourism and conservation more broadly.

However, in the current case study, it was the LFMC's responsibility to prove itself as a viable entity in the face of extractive industries. To establish this effectively, it was necessary to redefine land use practices by renting "pristine" lands from the local populace for the purposes of ecotourism ventures—these lands are far more accessible than the forested Crown Lands characterized by thick and virtually impenetrable bush. While this process accords with those outlined in the Forestry Act of 1996, whereby afforestation on private lands was promoted in Jamaica, it presents a unique set of challenges concerning the establishment of nature parks among both landowning and landless small-scale farmers, in that the development of such nature parks puts small-scale farmers' stake to their lands and sense of autonomy that comes with landownership in Jamaica into question.

"So long as surplus labor is manifested in agricultural commodities, economic and political power is closely tied to landownership" (Smith 2007). This presents a precarious set of circumstances concerning Cockpit Country in that the land in question was not owned by an elite class, but a community member of Bunkers Hill who was forced to relinquish his position in the LFMC due to his questioning of the language and process of sustainable development, but still TNC and FDJ attempted to appropriate the land in question toward the ends of the LFMC. Given these circumstances, the development process proposed in Bunkers Hill was likely to yield uneven outcomes among community members, particularly those with land holdings. However, the strategy employed in the current case certainly takes on a distinct tone of inequity concerning property ownership and use, in that those who were able to qualify their land rights were pressured to use their lands for the gain of others, particularly TNC. Historically speaking, with respect to development initiatives, the GoJ has been instrumental in stripping Jamaican people of their property rights. However, The GoJ would struggle to attain the land by any existing legal act, as the venture involved neither mineral extraction nor heritage preservation. However, the property in question presented a landscape that was particularly conducive to the establishment of an ecotourism site in Bunkers Hill—rivers, trees, and open areas—at least according to TNC's tourism attractiveness index. With one of the ecotourism locations in the hands of local small-scale farmers, TNC and FDJ's establishment of "nature as an accumulation strategy" in Cockpit Country required them to essentially place a "stake" between the landowning LFMC members' rights to property and land use to provide a basis for the reconfiguration of the production of nature in the realization of exchange values on the global market place. This particular form of tourism is geared toward the more affluent members of Jamaican and international communities, who frequently garner their own wealth from development, seeking to "consume supposedly pristine nature so long as that nature remains pristine, undeveloped" (Smith 2007). There is an inherent contradiction in this process as it was deemed necessary to strip the land from the local farmers to provide attractions to an elite group that, for all intents and purposes, has participated in producing the conditions that brought about the establishment of the proposed nature preservation ventures.

8.6 Conclusion

This chapter provides several inputs concerning sustainable tourism, whereby my analysis focuses on people's experiences of conservation and development in Cockpit Country, Jamaica. These experiences are quite revealing with respect to developing an understanding of the nuanced ways in which people think about the relationship between nature and society in the context of sustainable tourism and development more broadly. Further, pairing this with an analysis of the collaboration of small-scale farmers in Cockpit Country, people from the international conservation arena, international development, and national agencies in the development of projects that were intended to promote alternative livelihoods for the rural poor while conserving Jamaica's forests provides a situated and contextual mode for understanding the problems and potentials of sustainable tourism and sustainable development more broadly in Jamaica.

Cockpit Country is a place where many people who practiced small-scale farming in Jamaica were pushed to the margins during the colonial period. With the more arable flat lands on the plains being dominated by wealthier landowners in the planter class, the landless population sought out spaces in these "backlands" where there was little interest on the part of the colonial government; that is, until these lands were appropriated for timber and fuelwood for the railroad and other industries in the late nineteenth and early twentieth centuries (Satchell and Sampson 2003), as well as for land resettlement initiatives on the part of bauxite companies from the 1960s. It is clear that access to arable land and resources is a pervasive theme throughout the history of Jamaica, yet small-scale farmers in this area have been continually pursued by conservation, development, and now sustainable tourism and development programs leveraging forest and land use policies that threaten their access to land and resources.

Projects such as those promoted by the CCLFMCs and the collaboration of actors with varying interests and power inputs have the potential to mask the "inequalities and cultural distinctions" that are inherent in sustainable tourism and development projects due to the focus on an economic rather than ecological rational (Banerjee 2003). While this has been well documented in sustainable development projects throughout the global south, I would like to extend this broad analysis, whereby, yes, these projects certainly masked "inequalities and cultural distinctions"; however, such cultural distinctions may be further understood through people's understanding of sustainable tourism and development, whereby this understanding is produced and reproduced in the context of people's material and conceptual understanding of nature and society (Douglas 2014). In the current case study, the production of Cockpit Country was grounded in the tenets of sustainable tourism as a means for promoting the ideals of conservation and development outlined in the 1996 Forestry Act of Jamaica. This resulted in the reproduction of the systemic inequities experienced by small-scale farmers and forest-fringe dwellers since colonial times.

In the current case study, inequitable power structures between small-scale farmers and broader forest-fringe community members and NGOs/INGOs were pervasive. These inequalities were embedded in the structure of the LFMCs, which were starkly reminiscent of a clientelistic form of governance that was a fundamental part of the Jamaican political system up to the turn of the century (Sives 2009)—a system characterized by structural inequities. This begs the question of how committees, though developed on democratic principles, that were intended to promote forest conservation through sustainable tourism might exist while reproducing such structural inequities. In essence, the reality of these organizations is that not only were they grounded in the inequities that had been experienced by small-scale farmers since colonial times, but also attempted to further reproduce the dominant neoliberal conservation program. As has been documented in the literature, such programs tout ideology and practice that stems from faraway places that are equal in geographical and conceptual distance in terms of addressing social and environmental inequities. Meanwhile, these programs continue to operate in a top-down manner, effectively marginalizing the voice of the people that they work with (Brosius et al. 1998; West 2006).

How was such a process presented and accepted, even by those that recognized the broad inequalities in such programs? I argue that the language of sustainable tourism and development presented a mirage of irresistible opportunities; this language offered highly desirable outcomes for small-scale farmers and broader community members in Cockpit Country. These included access to food, education, and alternatives to farming, particularly for people in younger generations. Yet this language was embedded within a certain fetishization of nature, whereby the organizations that produced this language were keen to appropriate nature for sustainable tourism and a host of cottage industries that catered to the dominant western notion of neoliberal conservation.

In conclusion, the new age of sustainability, one grounded in myriad sustainable industries, e.g. ecotourism and fair trade, has not escaped the "same old story," a story of broad scale systemic inequities produced and reproduced through top-down environmental governance intended to control access to resources. This begs the question of how to move forward given the failure of sustainable industries to escape the dominant development paradigm that Escobar (1996) spoke of. Given this, if we begin to consider the nuanced ways in which nature is produced, as set forth by Smith (2008), it may be possible to identify the contradictions that tend to be inherent in sustainable tourism and the like before implementing programs that are more concerned with consuming nature (Smith 2007) than producing said intent. Furthermore, it will be instructive to include community members in the process of assessing the problems and potentials of such programs in a more systematic manner than was employed in the current case study. One suggestion concerns Community-Based Participatory Action Research (CBPAR), whereby community members, academics, and a broad range of stakeholders work collaboratively to identify issues, develop methods for investigating the issues, collect and analyze data, and determine how to move forward with addressing such issues through a more situated understanding of social, political, and environmental context (Cahill 2004; Minkler 2000). A final suggestion includes participatory program evaluation, which involves stakeholders in evaluating what works, what does not work, and how to move forward with program development (Baker and Sabo 2004). With these theoretical and practice-based suggestions, it may be feasible to develop programs that are better grounded in community needs and practices. Then, when 'dem come, the people that are most affected by sustainable tourism may be empowered to develop practices based on local knowledge, knowledge produced and reproduced through lifetimes of practice.

References

Baker, A.M., and K.J. Sabo. 2004. *Participatory evaluation essentials: A guide for non-profit organizations and their evaluation partners*. Cambridge, MA: The Bruner Foundation.

Banerjee, S.B. 2003. Who sustains whose development? Sustainable development and the reinvention of nature. *Organization Studies* 24: 143–180. doi:10.1177/0170840603024001341.

Brosius, J.P., A.L. Tsing, and C. Zerner. 1998. Representing communities: Histories and politics of community-based natural resource management. *Society and Natural Resources* 11: 157–168. doi:10.1080/08941929809381069.

Cahill, C. 2004. Defying gravity? Raising consciousness through collective research. *Children's Geographies* 2: 273–286.

Carrier, J.G. 2003. Mind, gaze and engagement understanding the environment. *Journal of Material Culture* 8: 5–23. doi:10.1177/1359183503008001760.

Dixon, H. 2006. *Bauxite mining: A threat to Cockpit country*. Albert Town: Southern Trelawny Environmental Association.

Douglas, J.A. 2013. *In the Cockpit: The political ecology of integrated conservation and development in Cockpit country, Jamaica*. New York: City University of New York.

Douglas, J.A. 2014. What's political ecology got to do with tourism? *Tourism Geographies* 16: 8–13. doi:10.1080/14616688.2013.864324.

Escobar, A. 1996. Construction nature: Elements for a post-structuralist political ecology. *Futures* 28: 325–343. doi:10.1016/0016-3287(96)00011-0.

Escobar, A. 2008. *Territories of difference: Place, movements, life, redes*. Durham: Duke University Press.

Evelyn, O., and R. Camirand. 2003. Forest cover and deforestation in Jamaica: An analysis of forest cover estimates over time. *The International Forestry Review* 5: 354–363.

Hale, C.R. 2002. Does multiculturalism menace? Governance, cultural rights and the politics of identity in Guatemala. *Journal of Latin American Studies* 34: 485–524. doi:10.1017/S0022216X02006521.

Katz, C. 1998. Whose nature, whose culture? Private productions of space and the preservation of nature. In *Remaking reality: Nature at the millennium*, ed. B. Braun and N. Castree, 46–63. New York: Routledge.

Kellert, S.R., J.N. Mehta, S.A. Ebbin, and L.L. Lichtenfeld. 2000. Community natural resource management: Promise, rhetoric, and reality. *Society and Natural Resources* 13: 705–715.

Lundy, P. 1999. Fragmented community action or new social movement? A study of environmentalism in Jamaica. *International Sociology* 14: 83–102. doi:10.1177/0268580999014001005.

Massey, D.B. (ed.). 2005. *For space*. Thousand Oaks: Sage.

McGregor, D.F.M., D. Barker, and S. Lloyd-Evans. 1998. *Resource sustainability and Caribbean development*. Kingston: The Press, University of the West Indies.

Minkler, M. 2000. Using participatory action research to build healthy communities. *Public Health Reports* 115: 191.

Orlove, B.S., and S.B. Brush. 1996. Anthropology and the conservation of biodiversity. *Annual Review of Anthropology* 25: 329–352.

Polanyi-Levitt, K. 1991. *The origins and consequences of Jamaica's debt crisis: 1970–1990*. Mona: Consortium Graduate School of Social Sciences, University of the West Indies.

Robotham, D. 2005. *Culture, society, and economy: Bringing production back in*. Thousand Oaks: Sage.

Satchell, V.M., and C. Sampson. 2003. The rise and fall of railways in Jamaica, 1845–1975. *Journal of Transport History* 24: 1–21.

Sives, A. 2009. Electoral reform and good governance: The case of Jamaica. *Commonwealth and Comparative Politics* 47: 174–193. doi:10.1080/14662040902857818.

Smith, N. 2007. Nature as accumulation strategy. *Socialist Register* 43: 16–41.

Smith, N. 2008. *Uneven development: Nature, capital, and the production of space*, 3rd ed. Athens: University of Georgia Press.

The Forestry Department of the Ministry of Agriculture of Jamaica. 1996. *The Forest Act 1996*. Kingston: The Forestry Department of the Ministry of Agriculture of Jamaica.

Thomas, D.A. 2004. *Modern blackness: Nationalism, globalization, and the politics of culture in Jamaica*. Durham: Duke University Press.

TNC. 2006. *Cockpit country conservation action plan (a summary)*. Washington, DC: The Nature Conservancy.

TNC. 2007. *Cockpit country, Jamaica: Parks in peril end-of-project report*. Arlington: The Nature Conservancy Worldwide Office.

USAID. 2009. *Biodiversity conservation and forestry programs.*. Retrieved from http://oai.dtic.mil/oai/oai?verb=getRecord&metadataPrefix=html&identifier=ADA525030.

USAID. 2010. *USAID country assistance strategy (2010–2014), Jamaica*. Washington, DC: United States Association for International Development.

Weis, T. 2000. Beyond peasant deforestation: Environment and development in rural Jamaica. *Global Environmental Change* 10: 299–305.

Weis, T. 2006. The rise, fall and future of the Jamaican Peasantry. *Journal of Peasant Studies* 33: 61–88. doi:10.1080/03066150600624496.

West, P. 2006. *Conservation is our government now: The politics of ecology in Papua New Guinea*. Durham: Duke University Press.

Wimbush, A, 1935. *Forestry problems of Jamaica*. Kingston: The Forestry Department of the Ministry of Agriculture of Jamaica

World Bank. 1999. *Jamaica-Cockpit country conservation ... project*. Washington, DC: World Bank Publications.

Chapter 9
Understanding the Himalayan Townscape of Shimla Through Resident and Tourist Perception

Rajinder S. Jutla

Abstract This study focuses on the Himalayan townscape of Shimla, a popular tourist destination. It analyses the development of Shimla in terms of its history and Butler's life cycle resort model. It compares and contrasts current development practices through site interviews with residents and tourists. It examines and discusses the city's recent growth in terms of three pillars of sustainability: environmental, social and economic concerns. The paper provides planning recommendations for creating sustainable development for tourism. It concludes that for Shimla to continue as a viable tourist destination, it must have a comprehensive strategic plan encompassing social, political, and economic forces to protect its natural and cultural landscape. Planners must realize that tourism and the quality of the physical environment are intertwined. A failure to protect and enhance this environment will erode not only the visual character of Shimla's townscape but will also result in environmental degradation.

Keywords Indian tourism • Heritage tourism • Tourism in Himalayas • Indian cities • Environmental perception • Urban tourism

The Himalayan Mountains are an intrinsic part of Indian culture. They are of particular significance in Hinduism. According to Hindu mythology this mountain range is considered to be the center of the universe and the axis of the world. Many popular pilgrimage centers such as Badrinath, Kedarnath, and Hemkunt are located in the Himalayas and have attracted many visitors over time. In the last decade, the number of visitors have increased dramatically giving rise to a rapidly growing regional tourism industry.

In the summer months, thousands of Indian and foreign tourists flock there not only for pilgrimage but also to escape the hot Indian summer. In winter, many go to experience snowfall and to participate in winter sports. To meet the rising demand

---

R.S. Jutla (✉)
Missouri State University, Springfield, MO, USA
e-mail: RajinderJutla@Missouristate.edu

S.F. McCool, K. Bosak (eds.), *Reframing Sustainable Tourism*,
Environmental Challenges and Solutions 2, DOI 10.1007/978-94-017-7209-9_9

for tourist accommodation, a large number of facilities have been constructed over a short period of time. Himalayan communities have also experienced tremendous growth due to their own population pressure. This has led to a loss of townscape character which has also been accelerated by the fast change in technology, construction systems, and building styles.

It is difficult to arrive at a standard definition for townscape. Terms such as townscape, 'cityscape', 'farmscape', 'seascape' and 'humanscape' are derived from the word 'landscape'. For example, Cullen (1961) in The *Concise Townscape* and Burke (1976) in *Townscapes* coined the term to connote the visual and descriptive survey of a British town whereas Smailes (1955) in *Some Reflections on Geographical Description and Analysis of Townscapes* uses the term to mean an urban scene, a tract of landscape distinct from its rural surroundings. Price (1964) also analyzed the urban landscape of an Italian city in terms of street buildings, open spaces and walls. He points out that 'the (urban) landscape provides an expression of the city's workings and its past. It also provides the resident or visitor with an impression of the city (p. 242)'.

Gruen (1964) in *The Heart of Cities* evolved a series of terms from cityscape to describe the visual chaos of American cities. These are 'technoscape', 'transportationscape', 'suburbanscape', and 'subcityscape'. Gruen's use of these terms is a part of critical analysis of the urban environment. Blake (1964) in *God's Own Junkyard* used the terms 'townscape', 'carscape', and 'seascape' to describe the deterioration of urban America and visual chaos through powerful photographs. Whyte (1968) used the term 'townscape' to describe ways to improve the visual quality of cities. Gruen, Blake and Whyte's definitions suggest the importance of the visual quality of towns.

These studies show that the visual quality of townscape is an important component of townscape and landscape conservation as well as the quality of the natural and built environment. It is also an important component of local heritage which includes local cultural traditions and landmarks. Finally townscape and landscape cannot be enhanced without making broad economic policies an important part of the comprehensive development of any community and in particular Himalayan communities like Shimla, Kasauli, Solan, Mussoorie, and Nainital. Townscape is therefore treated as a comprehensive concept used to address all the developmental issues of Shimla in terms of the three pillars of sustainability: environmental, economic, and social aspects.

This study examines the fast growing Himalayan community of Shimla in terms of resident and tourist perception of its rapidly changing townscape character and developmental practices. Shimla, a town known for its scenic beauty is tucked away in mountains covered with tall majestic pine trees. It projects the image of a typical English town. These characteristics make the city and the region an attractive place for tourists. The study examines the physical attributes of the townscape of Shimla through interviews with residents and tourists. The physical attributes of a townscape are an important part of the extrinsic dimension and are relevant to this study.

This study is important because in recent years, Shimla has emerged as a major tourist destination. This increase stems partly from the political unrest and lack of safety for the last 20 years in the Indian state of Kashmir which in previous decades had been a major tourist destination in Northern India. The city of Shimla attracts not only Indian tourists from different states, but also international tourists. For these foreign nationals, a visit to Shimla is often the first stage of a trek through the Himalayas. There is also an increase in weekend tourists from neighboring areas and states. Consequently, Shimla has become one of the most popular tourist destinations in Northern India.

The relationship between tourism and the physical environment has been recognized by tourism planners worldwide. Many world renowned cities, such as Salzburg in Germany, Innsbruck in Austria, and Banff in Canada are widely acclaimed and popular among tourists for their physical setting and townscape quality. In the words of Tringano (1984, p. 20) 'tourism and environment are inseparable.' Thus, environment, which attracts tourists, needs to be protected. This study of townscape character dealt with the urban environment which consists of both the cultural and natural elements of the city. The cultural landscape includes landmarks, unique historic buildings, and public spaces, and form the backbone of the urban fabric. Lynch (1996) identified landmarks as important to cities because they provide them with a visual identity and are therefore the urban signatures by which a city is remembered. The natural environment includes the topographic character, wooded areas or water bodies. Natural settings provide a backdrop for the city and play an important role in lending character to well known cities. Krippendorf (1997) clearly articulated that landscape quality is the capital in tourism and so must be managed properly. Tourism planners need to realize that it is the quality of the urban environment which attracts tourists. The lack of planning and management of the urban landscape will result not only in loss of visual aesthetics but also environmental degradation which may cause natural disaster and safety concerns.

The development of Shimla can be explained through Butler's life cycle resort model. This model provides an understanding of tourism development in terms of its past processes and plans for future development. Butler's life cycle resort model proposes six stages of development in tourist destination areas based on the number of visitors over time (Butler 1980). The six stages of development are exploration, involvement, development, consolidation, stagnation, and decline/rejuvenation. The exploration state is the initial stage of the tourism industry. This stage involves the arrival of a small number of tourists at a destination with limited tourism facilities. This stage existed before 1947 when India was a British colony. At that time about 50,000 British officers and rich Indians visited Shimla annually. After India's independence from the late 1940s to the late 1980s, tourist arrival to Shimla increased to approximately 250,000 annually; this period can be described as the stage of involvement. During this time the city started celebrating the Shimla Summer Festival which was marketed through national and various regional newspapers. A number of Bollywood movies were also shot in Shimla making the city popular among Indians. The number of tourists continuously increased. In late

1980s, an average of 350,000 tourists visited Shimla annually. The period 1980–2000 marks the development stage, the third stage of Butler's model. There were 50 hotels at the end of the 1970s but by the early 1980s, this increased to 75. *The Tribune*, a major regional newspaper, reports tremendous growth in Shimla's hotel industry, from 98 in 1990 to 210 in 1998 (Singh 2000).

In the early 1990s the annual tourist arrival increased to 500,000 and by the late 1990s, this number almost doubled to 900,000 (Singh 2000). During this time, residents felt the impact of tourism on their quality of life and some tension developed between the two groups. Property values and house rents increased and there was a shortage of drinking water. In terms of Butler's resort cycle model, Shimla, at that point, had only experienced the first three of the six stages of tourism development.

Tourism is still growing in Shimla and the city has entered the fourth stage of consolidation. This stage arrives when tourism develops into a major industry. The number of visitors has tremendously increased. Nowadays, over 1.25 million tourists regularly visit the city and its neighboring areas throughout the year. This is in part due to well marketed activities like festivals and events to showcase the culture and traditions of the region. *The Times of India* reports that there are more than 85,000 vehicles on the roads of the city every day. This number increases dramatically during the summer tourist season (Dhaleta 2013).

The fifth stage of stagnation is marked when tourism reaches its peak and is not managed properly. In this stage the place can no longer support the volume of its visitors and starts to lose its popularity as a viable tourist destination. The final stage of decline sets in when tourist arrival starts to spiral downwards. Tourism planners need to develop long term measures to prevent these last two stages from occurring.

A number of studies done in the past include social impact studies in terms of factors, such as, residents involvement with tourists, crime, economic activities, proximity to tourism activity areas, and traffic volume. Social impact studies explicitly state that both the cultural and natural environment are shaped according to people's perception, attitudes, opinion, and behavior.

An earlier study by Doxey (1975) concludes that as the number of tourists increases, the host resident population reacts in a hostile manner towards them. Butler's life cycle resort model (1980) and Doxey (1975) have provided a framework for understanding tourism development and residents' attitudes. According to their research, there is a strong consensus among the reactions of residents to tourism. In other words there is homogeneity of residents' perception of the impact of tourism.

Although a number of researchers have examined the reactions of host resident population to tourists, relatively few have focused on the reactions of residents and tourists to the changes in the townscapes brought about by tourism development. Kavallins and Pizam (1994), Caneday and Zeiger (1991), Lawson et al. (1998), Jurowski et al. (1997), Ryan et al. (1998), and Allen et al. (1988) have also conducted studies in the areas of perception and attitudes of tourists and residents. These studies examined issues of tourism, recreational development, and tourist destinations in terms of social, economic, and environmental aspects. However these studies focused on well developed resorts and urban destinations in the developed world. Relatively little attention has been paid to Himalayan destinations such as Shimla. Although Singh (1985, 1989a, b, 1991) and Singh and Kaur (1985) have completed a number of studies in the context of the Himalayan region, they focused on non urban environmental aspects and sustainable development of the ecologically sensitive region of the Himalayas. According to Page (1995) and Law (1996), very little has been done to explore issues and problems related to the urban planning and design aspects of tourism. Also, hardly any research has been done to explore resident and tourist response to the visual character of Himalayan urban communities.

This study makes not only a contribution to urban tourism literature but also provides important input for the development and management of hill town communities based on resident and tourist perception. It is also significant because very little has been done on urban mountain communities. Hardly any work has been done on Shimla and its surrounding communities in terms of urban design aspects based on the attitudes and perception of residents and tourists. A number of local residents, local officials as well as tourists show deep concern for the changing townscape of Shimla. Both residents and tourists fear that the city is losing its character (Kanwar 1997; Lohumi 1999). Today the planners of Shimla face a dilemma. Some strongly believe in preserving the townscape character of the city while others support its growth according to present social, economic and political forces. Ascertaining resident and tourist reactions is therefore an important contribution to resolving such a dilemma not only for Shimla but for many other destinations facing similar issues of over-development.

The paper has three objectives. The first is to examine the historical development of Shimla. The second is to examine its favorite urban spaces, townscape character and pattern of current development in terms of resident and tourist perception. The paper attempts to address the development of Shimla in terms of townscape characteristics and to investigate how new buildings relate to their immediate surroundings and landscape. The third objective is to examine sustainable development and to make recommendations pertaining to present and future developmental issues of Shimla.

The first section of the paper provides a brief historical account of the city. The second section deals with the research methodology, analysis and a discussion of the resident and tourist response to the townscape character and pattern of development. The third section discusses sustainable development with recommendations for townscape enhancement and management.

9.1 The Site Description

Shimla is about 225 miles north of the Indian capital of New Delhi. It is connected by air, rail, and a network of roads to Delhi via Kalka, a town at the foot of the Himalayan mountain range. The journey from Kalka to Shimla is virtually an uphill one. The scenery is breathtaking and the route passes through intact cedar, pine and oak forests. The connection by train was made by a narrow-gauge rail line link, covering a distance of 56 miles through over a hundred tunnels through the mountains, on a journey which provides picturesque views of step farming villages, wooden houses and sloping roofs.

Originally, Shimla was planned for a population of 30,000–40,000 but the present permanent population, with the inclusion of peripheral villages has crossed 200,000. It is located in the Shivalik range of the Himalayas in Northern India and has become a very popular tourist destination in recent times. Shimla was carved out of mountains densely covered by mature pine forests, at an elevation of approximately 7500 ft above sea level. These majestic trees are the pride of the Himalayan region and they greatly contribute to its scenic quality (Singh 2000). Historically, the British colonials created Shimla as their summer capital because of its unique setting and a climate reminiscent of England. The hill station not only was a haven from the hot Indian summer but its winter snowfall allowed for recreational activities such as skiing and skating. To create a sense of familiarity, the British recreated an English townscape in Shimla. The image of the city was projected to revive memories of England. This is reflected today in a number of buildings such as the municipal corporation, state tourism information bureau, Church of Christ, and the State Bank of India.

9.2 Historical Background

Shimla was discovered by British army officers in the early part of the nineteenth century. At that time, it was a small hamlet, known as the abode of the local Hindu goddess, *Shamala.* In 1827, Lord Amherst the Governor General of British India built his summer residence there. After this it did not take very long for Shimla to become the favorite destination of young British officers. Far away from the conservative environment of Victorian England, Shimla was a place of fun, fancy dress balls, horse riding, and picnics. It was also known as the place where unmarried British women went to find a suitable match (Lohumi 1999).

The town grew quickly after 1864 when the Viceroy at the time, Lord John Lawrence made it the official summer capital of British India. The city was developed over seven spurs around an irregular shaped Ridge. Many important buildings were built in the last part of the nineteenth century. The Cathedral of Saint Michael and Joseph was built in 1882 and Rippon Hospital was built in 1885. The Gaiety Theater which was modeled after the Royal Albert Hall of London was built in 1887. The most well known and architecturally significant building which was completed in 1888, is the Viceregal Lodge, the summer residence of the Viceroy. This building is presently known as the Indian Institute of Advanced Studies (Lohumi 1999).

The urban form of Shimla was based on segregation. This was an essential element in preserving the existing social structure with the British rulers at the top and the Indian population on the lower tier (Kanwar 1997). Shimla has two major shopping areas. The first, the Mall road, was created to serve the needs of the colonials while the second, the Lower Bazaar, catered to the Indians. These two commercial areas are located in close proximity in the heart of Shimla, at two different levels. The Mall road is located at the upper level and the Lower Bazaar at the lower level. Both these areas have mixed land use which includes shopping, office, and residential spaces. English type shops, catering to the elite, were built on the Mall road. Also, a Swiss hotel chain opened the upscale Cecil Hotel. Nowadays the Mall is mainly frequented by wealthy Indians and tourists while the Lower Bazaar continues to meet the needs of a large number of Shimla residents.

In the early part of the twentieth century, a color bar was instituted to prohibit Indians from the Mall and the Ridge. This was lifted during the First World War when Indian men were recruited into the British armed forces. The wives of these Indian army officers started appearing on the Mall as well as at the Gaiety, a theatre frequented by the British residents. Elite schools for the children of the British and wealthy Indians were built at this time. Examples are Bishop Cotton and St. Edward's for boys and Loreto Convent and Auckland House for girls. These schools were run by missionaries from Great Britain. After the end of World War I, there was hardly any development in Shimla and the city continued to act as the summer capital of India up to its independence in August 1947 (Lohumi 1999).

At the time of India's independence, Lahore, the capital of the adjoining province of Punjab was transferred to Pakistan. As a result, Shimla served as the temporary capital for the state of Punjab. In 1956, the capital of Punjab was transferred to Chandigarh, a modern city designed by the world renowned architect, Le Corbusier. At this time, many government offices moved to Chandigarh. Another turning point for Shimla came in 1971 when the city became the state capital for Himachal Pradesh. It continues today to serve in this capacity. Another change came about in the late 1980s when the spelling of Shimla was changed from the British name of 'Simla' to Shimla, in accordance with the pronunciation of the local residents.

9.3 Research Methodology

A questionnaire based survey was administered at the site during the summer tourist season to examine the current issues and problems of the townscape character and development practices of Shimla in terms of the perception of tourists and residents. Fifty residents and 50 tourists were interviewed face-to-face in the heart of the city, the Mall Road and the Ridge.

These areas were deliberately selected because they are major activity areas of the city. Many offices, popular restaurants, shops, and emporiums selling specialty items of the region are located in the Mall and on the Ridge. A good mix of residents and tourists from all over the city are represented daily and since the survey was conducted during the peak of the tourist season, there were many families from different parts of India and abroad. The sample included residents older than 20 years of age who have been living in Shimla for at least 10 years. The 10 year residential requirement was included to ensure respondents' familiarity with the city. To avoid gender bias, the sample was stratified to include 50 % male and 50 % female. Also, only one person per household was eligible to participate in the survey. The tourist population was defined as all those who had visited Shimla at least once before to ensure that they had some familiarity with the place.

The questionnaire was divided into three sections. The first section contained questions pertaining to demographic information such as age, place of residence, and profession. This was identical for both groups. The second section of the questionnaire contained questions targeted to each group. Residents were asked about their length of residence in Shimla and about their satisfaction with the city. Tourists were questioned about their reasons for visiting Shimla, their accommodation, their place of residence, and vacation period. The third section was again identical for residents and tourists; it examined townscape character, the pattern of development, the most favorite and the least favorite area.

9.4 Analysis and Discussion

Age was classified into three categories: the first, between 20 and 34; the second, between 35 and 49, and the third, 50 years and over. Data compilation shows that residents interviewed for the study included 24 % between 20 and 34, 36 % between 35 and 49 years, and the remaining 40 % over 50 years of age (see Table 9.1). This selected sample of residents represented a range of occupations including military personnel, health occupations, educators, businessmen, civil servants and agriculturists (orchard owners).

Eighty percent of the residents who were questioned were born in Shimla and lived there all their lives. When asked whether they liked living in Shimla, 90 % responded, in the affirmative, they were satisfied with living in the city. They stated that it was quiet, safe, and pollution free as opposed to the hustle and bustle of big Indian cities. The remaining 10 % of the residents said that Shimla was boring and unexciting and that they would like to move to a bigger city in the future. When asked about the influx of tourists to Shimla, 52 % of the residents had a strong positive attitude and felt that tourism should be further encouraged. Twenty percent were somewhat positive while the remaining 28 % had a negative attitude and did not think tourism should be encouraged. Responses to the question on the impact of tourism were also mixed. Forty percent of the residents strongly believed that tourism is making a negative impact. Twenty percent somewhat agreed with this

Table 9.1 Age structure of residents and tourists

Age categories	Residents (%)	Tourists (%)
20–34	24	16
35–49	36	20
50+	40	64

Table 9.2 Tourists' reasons for selecting Shimla as a tourist destination

Criteria	Percentages (%)
Cool climate	50
Fresh air	10
Location and proximity	16
Unique character	12
Magnificent scenery	10
Cleanliness	2

while the remaining 40 % did not believe this. Residents commented that the cost of living shoots up during the tourist season and that the city cannot meet the increased demand for water and electricity.

Data analysis shows that tourists interviewed for the study included 16 % between 20 and 34 years old, 20 % between 35 and 49 years, and the remaining 64 %, 50 years and over (see Table 9.1). It also shows that 60 % of the tourists visited Shimla on more than two occasions. In response to the question about the reasons for selecting Shimla as their tourist destination, 60 % of the tourists responded that they came to Shimla because of its cool climate and fresh air. Sixteen percent said that they selected Shimla because of its location and proximity to their place of residence. The remaining 24 % chose Shimla over other hill resorts because of its unique character, magnificent scenery, and cleanliness (see Table 9.2).

Eighty percent of the tourists surveyed said that they would definitely return to Shimla. The other 20 % responded that they would like to visit a new destination. Eighty percent of the tourists stayed in hotels while the remaining 20 % stayed with relatives and friends. It was interesting to note that tourists who stayed in hotels spent one to a maximum of 2 weeks in Shimla while those who stayed with relatives spent more than 2 weeks. Five percent of these tourists also visited Shimla in the winter season to experience the snowfall and enjoy ice-skating.

9.4.1 *Favorite Urban Spaces*

Both residents and tourists were provided with a list of distinctive areas in Shimla. They were asked to identify their most favorite and least favorite parts of the city. Both groups were asked to identify the criteria which attracted them to these places and select any number that was appropriate for them. Eighty percent of residents described the Mall Road and the Ridge as their most favorite part of Shimla. This was because of location, a traffic free pedestrian friendly atmosphere, and people watching activities (see Table 9.3).

Ninety percent of the tourists also identify The Mall and The Ridge as their most favorite place. Tourists indicated that the Ridge is particularly important because it is the place where the summer festival is celebrated and it is where Shimla's well known landmark, the Christ Church, is located. The groups surveyed

Table 9.3 Criteria for the most favorite area

Criteria	Residents (%)	Tourists (%)
Central location	90	90
Pedestrian friendly	75	85
Restaurants	45	80
Movie theaters	60	65
Views	60	80
Activities for children	70	90
People watching	70	80
Shopping facilities	60	80

further elaborated that they always associate Shimla with the image of the church on the Ridge. Some tourists commented that the Bollywood movies shot in Shimla always included shots of the church. This illustrates the extent to which the city's image is promoted by Bollywood movies. The popularity of this spot is marked by horse riding activities for children. The Ridge is also connected to the Lakkar Bazaar which sells the specialty wooden items sought after by tourists.

Residents and tourists described the Mall Road and the Ridge as the hub of the town, as crowded, fun and pleasant places. They are the two important pedestrian oriented urban spaces of Shimla. They are centers of activity interlinked by important pathways. They generate a rich social life by attracting a large number of people. Both groups stated that they went there to meet their friends, stroll, wear their fashionable clothes and jewelry, and watch other people (Jutla 2000). People watching in urban plazas has been identified as a popular activity (Whyte 1980).

9.4.2 Least Favorite Urban Spaces

Sixty five percent of the residents identified the two Bus Terminals areas which are located at different points along the Circular Road as their least favorite places in Shimla. Fifteen percent of the residents could not identify any place and there was no consensus among the remaining 20 %. The Bus Terminal areas were again cited by 50 % of the tourists as their least favorite part of the city. Thirty percent of the tourists also identified the Lower Bazaar and Sabzi Mandi as their least favorite places in the city. There was no consensus however among the other 20 %. The least favorite areas were described as congested, polluted, unsafe, and unattractive (see Table 9.4). The least preferred areas identified by the majority of tourists were located in the central part of the city as this is where they spent most of their vacation time. Residents, on the other hand, identified areas all over the city since they had more knowledge of the place.

Table 9.4 Criteria for the least favorite area

Criteria	Residents (%)	Tourists (%)
Congested	84	90
Polluted	80	84
Unsafe	90	92
Unattractive	75	80

Table 9.5 Assessment of townscape character: new development

Townscape assessment criteria	Residents' sample mean	Tourists' sample mean
Appropriateness to the place	2.1	2.4
Appropriateness of building style	2.3	3.1
Appropriateness of construction system	1.8	2.9
Scale, color and texture of new building material	2.6	3.3
Overall mean ratings of townscape character	2.2	2.9

9.4.3 *Townscape Character*

The townscape character of Shimla has eroded since India's independence from Britain. When questioned about the character of Shimla, the majority of the residents and tourists were concerned about the changing character of the city. Residents and tourists were asked to rate the new development in terms of its appropriateness to the surroundings. They were assessing this in terms of construction method, building material, textures, scale and style. They were shown photographs of development in Shimla to explain these aspects. Since new development is seen in almost all corners of the city, it is difficult to ignore. Tourists who were visiting Shimla after a long period were very quick to observe the changes. They were asked to rate these questions on a 1–7 scale where 1 represented inappropriate and 7, highly appropriate. Table 9.5 summarizes resident and tourist ratings of townscape character. The overall mean ratings of these questions represented the townscape character. The average rating for all residents was 2.2 and for tourists it was 2.9. These ratings reflect a deterioration of Shimla's townscape.

To study the significant difference between the perception of development of both groups, the mean ratings and overall mean ratings of both residents and tourists were run through the z test for the two independent groups on Excel. The null hypothesis for each criterion which states that there was no significant difference between the mean ratings of residents ($n = 50$) and tourists ($n = 50$) failed to be rejected at $\alpha = 0.05$ significance level. This shows that there were no significant differences between the perception of both groups. There is no doubt that the mean rating for the residents were found to be lower than that of tourists. This shows that residents were more dissatisfied with the new buildings and land development schemes as compared to the tourists. However, both these resulting mean ratings were below the middle value of 4 on the measurement scale of perception on a seven point scale where 1 represented inappropriate and 7, highly appropriate. These ratings reflect that both the groups noticed a deterioration of Shimla's townscape character.

Eighty five percent of the residents were concerned about the inappropriate pattern of recent development which was described as characterless and ugly to look at. The residents were also very nostalgic for the old Shimla, its colonial character and cleanliness, particularly those who were born and raised in the city. They referred to the 1960s and 1970s as the 'good old days.' Residents also reported about the changing face of the Mall road due to the construction of a number of concrete buildings which do not lend to the local character of the older existing buildings. New buildings are not architecturally sympathetic to the existing buildings. Many of the old colonial buildings have been consumed by fire and residents complained that the remaining ones are not properly maintained (Jutla 2000). Shimla which has become very congested and overcrowded over the past 25–30 years is turning into a 'concrete jungle,' far from its original natural environment. Only 15 % of the residents thought that the change was positive in terms of cable TV, cellular phones, increase in public transportation, and private auto ownership. Eighty percent of the tourists also thought that the change was negative. The tourists who visited Shimla in 1970s and early 1980s romanticized the old image of the city. The remaining 20 % did not notice any significant change or thought that the change was unnoticeable. Twenty percent of tourists stated that the change was positive in terms of better train travel facilities, quality of hotels and taxi service.

9.5 Current Pattern of Development

In response to the question about the current practices of physical development, 87 % of residents and 80 % of tourists responded negatively to the way Shimla was developing. Both groups agreed that the new development is haphazard, and ugly, resulting in overcrowding, and congestion. This is happening because of urbanization, population growth, and modernity.

Lohumi (1999) reports that a number of high-rise apartment buildings are mushrooming all over the city. They have even been constructed on lawns and tennis courts of old buildings. The Tenancy and Land Reform Act, which debars non agriculturists from buying land in the rural area is responsible for the congestion and the concentration of construction activity in the city. The increased demand for housing has encouraged property owners to construct multi story buildings, occupying one floor, and selling the others to pay for the cost of the land and construction. Sharma (2001) reports that in the past many people who lived in single or double story houses rented their additional accommodation.

In the last 15–20 years, the population pressure has added more cars and concrete unplanned buildings in Shimla. This is the result of the recent economic prosperity of the region. Many buildings were constructed rapidly and in many cases, without ecological, aesthetics and safety concerns. The residents believe that they can build almost anything with concrete pillars but this has proved to be disastrous since many of these high-rise structures collapse because of landslides and inappropriate construction on unstable terrain.

Forty percent of the resident sample has been living in Shimla for more than 25 years. They were also nostalgic for some of the interesting streetscape details from the colonial past. The same feeling was also expressed by 20 % of the tourists, who had visited Shimla on several occasions before 1980s. For example, many residents talked about the elegant design of the old red post boxes, decorative wrought iron fences, manhole covers on roads and also unique lion-headed public water taps on the roadsides. The people surveyed wondered why these details have disappeared and why similar details are not utilized at present.

Townscapes are changing rapidly because of modernity and new life styles. With the popularity of cellular phones, very tall steel communication towers are erected on hills. Although necessary, they can be located at appropriate places to minimize the visual impact. Townscapes are also changing because of the invention of new materials and construction systems. Deforestation has made timber scarce, expensive and not easily available. As a result concrete modular systems are used extensively for construction. Also the popularity of cable and satellite television has resulted in the crisscrossing of cables throughout the city along with satellite dishes on rooftops. Cable TV is operated by private operators and there are absolutely no guidelines for them. Although strict bylaws do not have to be enacted to freeze the townscape character, it is necessary for the new development to be harmoniously linked with the city's history and its future development.

Both residents and tourists agreed that the new development is not visually pleasing and lacks creativity. It is also inappropriate in terms of building material, scale, and style. The new development is fragmented and do not follow the pattern of existing development. Both groups agreed that there is overbuilding in the city resulting in congestion and safety hazards. New buildings are separated by very little space. Both residents and tourists indicated a strong preference for more traditional building styles.

The two groups also indicated that there are no well defined and strict guidelines for the new development. It was also pointed out that the development approval process was unfair and corrupted. According to Sharma (2000), many areas in the vicinity are not covered under the Town and Country Planning Act and the local officials are not armed with effective land use laws.

Politicians and planners need to realize that environmental conservation and protection is an essential ingredient of the tourism industry. In order to preserve and conserve the natural landscape, development should respect nature and should be sympathetic to the existing landscape. There is an urgent need to enact and enforce design guidelines for new development.

9.6 Sustainable Development

Sustainable development can be described as a development which satisfies the needs of the resident population and the present influx of tourists without compromising the needs of future generations. The concept of sustainability is rooted in a system approach which requires one to think in terms of the interconnectedness of various systems. In this study, sustainability will be examined in terms of its three important pillars of Environmental, Social and Economic concerns.

9.6.1 Environmental Concerns

Townscape can be seen as a product of complex, interconnected, environmental factors. These factors were first recorded 2400 years ago in Greek history. Plato, the Greek philosopher wrote about soil erosion and deforestation caused by overgrazing and tree felling in the hills for fuel (Middleton and Hawkins 1998, p. 16). Today they seem to be just as relevant. The development and expansion of Shimla over the last 20 years, for example, have led to the destruction of forest cover. It was this forest cover that earned the city the title of 'Queen of Hills.' During the last two decades a large number of trees were cut or cleared for construction with little or no effort made to replace them. No new trees were planted although the local corporation would symbolically perform the ritual of tree planting on occasion. People use illegal methods to cut trees in order to make way for the construction of new buildings and parking lots for vehicles. This has led to landslides and land erosion. In some areas pollutants from garbage are leeching into the soil causing a dramatic transformation.

Population increase and the rise of the middle class have also put pressure on land development. Mountain slopes previously considered inappropriate for development are now being exploited for the construction of multistory buildings using concrete pillars. This type of heavy construction though costly and hazardous have become commonplace. Many residents and tourists expressed safety concerns about its escalation since construction of this type and the resulting deforestation have caused landslides and soil erosion.

The issues of landslides are not adequately dealt with. The major landslides of 1971 caused a serious threat to the city water reservoir beneath the Ridge. In May 2000, a building collapsed following a landslide which was triggered by the haphazard cutting of the hill below for the construction of a hotel. Landslides in 2001 have also caused a lot of damage and it took a long time for the city to repair these roads. The area around Lakkar Bazaar and the Grand Hotel has been declared landslide-prone but buildings are still being constructed despite a ban by the government. Indiscriminate digging of hill slopes for construction coupled with the weathering of rocks, has led to reports that Shimla is sinking at several places. For example, the road connecting the Ridge with Lakkar Bazaar has sunk about 2 ft below road level in the last few years. Shop owners have been forced to construct steps in order to connect the road to their shops. The widening of roads, inappropriate construction on hills, overgrazing on hillsides tree felling, dynamiting, and the removal of vegetation have all contributed to the destruction of the natural landscape. This can also be observed in the rural communities along the Shimla – Kalka highway.

The city of Shimla which was known for its enduring imagery of majestic pine forests is being replaced by a concrete jungle with monotonous ill planned and poorly designed buildings. These trees are the pride of the Himalayan region and greatly contribute to its scenic quality. They are an important economic resource of the region. The major selling point of the area is the scenic resource of the area and its climate.

There is therefore an immediate need to preserve the natural landscape of Shimla and the surrounding area. Government agencies must develop a comprehensive approach to the townscape management of the entire region. It is important to identify the unstable slopes in order to establish no construction zones. Local authorities seem to be ignoring these problems because of political pressure. They must address the issues of sinking roads and buildings by setting standards for design and construction based on principles of environmental planning and design.

The environmental disaster of 2013 which was caused by heavy rainfall in the Himalayan state of Uttrakhand can serve as a lesson to other Himalayan communities. The heavy loss of human life and property damage was the result of a poorly planned infrastructure where very little attention was paid to environmental concerns. *The Times of India* reported that people in the region built the cheapest and quickest motels, restaurants, and roadside kiosks to make profits from the growing number of tourists. Trees were cleared on many hill sides for the construction of newly erected buildings. Roads were widened to accommodate a large volume of tourist traffic into the area. The sensitivity of the mountains' ecosystems was sadly overlooked (Sethi 2013).

Deforestation in the Shimla region has also affected climatic conditions. The amount of snowfall has decreased substantially over the last 20 years. This was a major draw for tourists since many of them would come from the plains to enjoy the snowfall in Shimla and to participate in winter activities like ice skating and skiing. The decrease in precipitation has in recent years, created a water shortage in the region. Also in 1970s and 1980s people did not need the use of fans or air conditioning in the summer months but again, because of deforestation, the summer temperatures of the region have increased. This temperature change may also be attributed to the growing number of automobiles in the area. The increase in carbon monoxide and carbon dioxide emission into the air has led to the creation of an ozone layer thus causing a greenhouse effect.

9.6.2 Social Concerns

Townscape planning and management should address the social concerns of both residents and tourists. Social sustainability should be concerned with enhancing the quality of life within the community.

Shimla and its surrounding areas provide a unique setting for social, cultural, and religious activities. These include the Shimla summer festival, hiking, picnicking, bird watching, visiting the famous Hindu temple at the top of the highest mountain peak, and experiencing the natural beauty of mountains draped with pine trees.

It is important to properly maintain the central areas of Shimla. The Mall and the Ridge generate a rich social life for both residents and tourists. As described earlier, these areas are very lively and full with people. Some tourism activities cater to diverse age groups. Younger children and teenagers enjoy horse riding and many young adults and older people enjoy strolling and socializing with friends at the various restaurants on the Mall and the Ridge. These areas are virtually pedestrian zones since only emergency vehicles are allowed. During the peak season the demand for benches exceeds supply.

The tourism industry should not only cater to a diversity of people in terms of age groups and also to the domestic tourists from various states of India as well as international tourists. Tourists should be introduced to local culture and traditions. This can be done by developing certain villages where people can experience life in a Himachal village, visit fruit orchards and learn about the regional handicrafts. Similarly, tourists should also be introduced to Tibetan culture since a large number of Tibetans have settled in the area since the 1960s. These are the people who fled their homeland after its takeover by China. Tourists can be introduced to the traditional Tibetan way of life, the culture and cuisine. The Tibetans already play an important role in the tourism retail sector.

Many residents think that tourism has many negative aspects. They complain that it is eroding the uniqueness or identity of place. To meet the demand of a large influx of tourists, buildings are being constructed with no regard for context. This needs to be addressed soon in order to preserve the architectural heritage of Shimla which includes a number of unique buildings constructed during the British rule. Many of them were built in the Tudor style with stone quarried in the area. These buildings need to be preserved because they lend diversity to the townscape character. Many of the areas which are described as favorite spots by both residents and tourists include these buildings. To preserve the sense of place or the spatial character of the area, emphasis should not be given to preserving buildings in isolation but they must be analyzed in the context of the entire streetscape or public space in terms of scale, style and material.

The BBC reports that in recent times some British tourists are coming to Shimla and its surrounding areas to visit graveyards where their ancestors were buried during the colonial era. A large number of British people had relatives who served in India as governmental officials, soldiers, shopkeepers, traders, tea planters, forest officials, teachers, and missionaries etc. These tourists visit India to find the gravesites of their ancestors and also to see where they lived and worked. Many of the graveyards are located in and around Shimla since this was the summer capital during British rule (Biswas 2006).

9.6.3 Economic Concerns

Successful townscape planning and design cannot be done without giving serious consideration to economic issues. Traditional economic theory does not take into consideration the depletion and degradation of the natural environment and the importance of ecosystems which are generally regarded as free public goods. Landscape and townscape aesthetics is a valuable economic resource. Vegetation on the mountains plays a significant role in reducing the impact of natural disasters. For example, it helps in protecting against erosion, landslides and local flooding. It protects people and property from rock fall, and the water-holding capability of mountain vegetation reduces water runoff. The value of mountain ecosystems in protecting against natural disaster can be deduced from the high economic and social costs of the recent cloud burst in the Himalayan region of Uttarakhand.

Shimla and the surrounding Himalayan region are known for their fruit orchards, nuts, potato crops and medicinal herbal plants. These satisfy not only local needs but are also exported out of state. They can therefore provide tremendous economic benefits to the local people. Medicinal herbal plants can not only benefit the local population but also global communities. This potential should be harnessed through the setting up of pharmaceutical plants. Local food products and beverages should also be promoted. In addition, tourism development projects should have at least 50 % local ownership to discourage economic leakages and ensuring that the profit stays locally. Since Shimla is also home to a state medical college, health tourism can be promoted nationally and internationally. This will help to provide employment for local people.

There is a dire need to educate the people living in the Shimla area on the environmental, social, and economic benefits in preserving the sensitive mountain ecosystem. Sustainable development should be done to conserve and preserve the natural and built environment in such a way that future generations can derive at least the same benefits.

9.7 Recommendations

Planning provides an important tool for creating sustainable development for tourism. Without planning it will be difficult to address complex aspects of environmental, social and economic issues. A fragmented, piece meal approach as happening now does more harm and fails to provide benefits to both host population and tourists. The following steps are recommended for the development of a sustainable tourism plan:

Step 1: Understand the Wider Context: Goals and objectives of a tourism plan should be developed in the wider context of the area in terms of its natural, cultural and economic aspects. One of its most important objectives should be to preserve the natural landscape and history of the place since they lend to its townscape character.

Step 2: Develop a Land Use Plan: A Land Use Plan is the most basic and effective tool available to local planners since it is within the jurisdiction of Shimla's local government. Land use controls should be applied to all kind of physical development plans of tourism, recreational, and housing projects from small to large scale. These should be developed in the visual context of a mountainous region. Planners should also bear in mind that since Shimla is located on mountain slopes, its development is highly visible from far off and from many vantage points.

The planning and designing of both tourism and non-tourism developments should be based on the unique characteristics of the different areas of the town. It is recommended that various maps of topography, slope analysis, existing buildings, utilities, existing vegetation, and land slide prone areas should be prepared for different parts of the city and they should be available to developers and general public. Land use maps should identify dense mature vegetation which needs to be protected because it contributes to the unique and visual quality of townscape.

In Shimla construction on a slope requires cut and fill on slopes and changing the shape of the hill or context. This disturbs the natural environment, existing drainage system, and often requires the removal of vegetation. As a result, before approving any development project, an Environmental Impact Assessment needs to be conducted for all tourism and non-tourism projects to ensure that they would not degrade the natural landscape and create hazard.

In recent years with the development of sophisticated technology, new tools of Geographic Information System and Remote Sensing are available to planners and designers. It is important to remember that tourism planning should be an important component of the overall or comprehensive planning of Shimla. The following steps should therefore be taken:

(i) *Conduct Landscape Aesthetics & Biodiversity Analysis*
Landscape analysis should be conducted of Shimla and its surrounding areas. Areas of scenic quality based on good views and angles and biodiversity should be identified. Strict laws should be enacted to protect the scenic resource and bio diversity for the enjoyment of both tourists and residents.

(ii) *Conduct Townscape Analysis of Cultural Heritage*
This should include all important historical buildings and landmarks. Their physical conditions and their immediate context should be examined. Strict building bylaws to enhance or preserve their context should also be put in

place otherwise the encroaching of new character less and ordinary buildings will soon erode the sense of place or character of the historical area.

Land use planning controls can definitely help in achieving the sustainable development of facilities for both residents and tourists. It may be necessary for local or state government to develop land in public interest to protect special areas of natural and cultural resources.

Step 3: Design in Context

New buildings should create visual linkages between existing and proposed buildings to create a cohesive effect for the wider context. The new building should enhance and strengthen the visual characteristics of townscape. Buildings design should be visually compatible to create an overall effect greater than its individual effect.

Once innovative building regulations are enacted and enforced by strict laws it would become easier to deal with new building development. Building regulation should be comprehensively developed and should be based on building heights, densities, floor area ratios, color, material texture, form and context or relationship to the site. For example, heavy construction in this mountainous earthquake prone region may not at all be suitable.

To create a sense of naturalness and to relieve the sense of congestion, development should not be allowed to cover the entire space or site. At least 40–50 % of the site area should be left for open space and greenery. This will help to soften the current image of Shimla as a "concrete jungle".

In Shimla, the core areas, such as the Mall and the Ridge should be designated as a heritage zone and thus earmarked for conservation. Strict guidelines should be developed to protect the sense of place or character of the area. Also pedestrian corridors from the Mall to Summer Hill, from the Mall to Chota Shimla, and from the Mall to Sanjauli should be given special attention. These are important pedestrian movement paths for both residents and tourists and as such should remain solely pedestrian except for the occasional emergency vehicle. It is important to keep in mind that all land use control and building regulation decisions are not made by the planners alone. Most of the time, these decisions are made by the politicians. Successful policies should be based on active input from both residents and tourists.

Politicians and planners must understand that it is Shimla's sense of place which attracts tourists. If this is lost then the city will lose its popularity as a popular tourist destination. Shimla has to develop its tourism as a part of its total comprehensive growth which would accommodate both increased population and modernity.

Step 4: Infrastructure Development: An important consideration in the development of infrastructure, specifically the building of new roads is to design and plan them according to existing topographic characteristics. Minimum change should be done to existing contours and slopes in order to avoid landslides and conserve the existing trees.

In recent years Shimla residents have experienced water shortage during the tourist season. This led to water being rationed with availability only 1–2 hours in morning and 1–2 hours in evening and sometimes only 2–3 days in a week. This situation has evoked ill feelings of residents towards tourists who are blamed for this.

Before initiating new tourism projects or plans, the carrying capacity of Shimla's resources should be first assessed. Then, after the implementation of the projects, the site conditions should be monitored so that negative impacted issues such as land sliding or land sinking could be addressed.

To address the current issue of traffic congestion, the carrying capacity of the existing road network should be examined. There is a need to upgrade this and a new traffic management plan should be formulated and implemented.

Step 5: Develop Social and Cultural activities

Tourists should be given an opportunity to experience the unique life style and culture of the local people of Himachal as well as the Tibetans living in the area. Shimla should also develop graveyard tourism for the British people who come there to visit the graves of their loved ones. Cemeteries should therefore be properly maintained and also information about the graves of the deceased British colonials should be computerized to provide online access to people around the world

Step 6: Environmental Awareness Education

To preserve and conserve the sense of place of Shimla and its landscape, environmental education and eco awareness should be made compulsory and should be an important component of secondary school education. Local population should also be made aware of the negative consequences of hasty profit driven projects. The chopping of trees without proper permission should be heavily fined and tried in court. At present fines are nominal and as a result many people prefer to go against the established system and guidelines and pay the penalties and fines.

The local government should encourage developers to use modern technology of solar and wind power to address the shortage of energy. Using solar energy to heat water for bathing in cooler climate can conserve energy and also help in preserving trees which are cut for fuel in the area.

Step 7: Encourage Citizen Participation

The local residents of Shimla should be encouraged to participate in tourism planning policies and their input should be an important basis for developmental policies. Identify the goals of the tourism policies and determine how participation can assist in meeting them. Increase awareness of nature, conservation and preservation through the initiation of public awareness projects.

Step 8: Recognize Citizen Contribution to Townscape and Landscape Enhancement

At local level people who take an active role in landscape preservation, historic building conservation, finding solutions to traffic congestion should be recognized and rewarded in some way. Local people should be made to realize the importance of tourism as a revenue generator which can be used to improve quality of life in Shimla and its surrounding areas.

There is an immediate need to have well planned development as a part of a comprehensive sustainable tourism policy for Shimla and other hill towns otherwise they will soon lose their sense of place and as a result the number of tourist arrivals and tourism revenue will decline. At this stage new development will go to unspoiled landscape and once again destruction of the landscape is repeated.

9.8 Conclusion

Tourism in the Himalayas has existed for a long time as a number of well-known Hindu pilgrimage centers are located there. When the British colonists established hill towns, such as Shimla, Mussoorie, and Darjeeling to escape the hot Indian summers, the number of tourists visiting the Himalayas increased over time. In 1864, Shimla became the summer capital city and administrative center of British rule in India. After that it attracted many visitors, and nowadays it has become a popular and rapidly growing tourist destination.

The study concludes that residents and tourists perceive that the city of Shimla is in danger of losing its unique townscape character as a result of its increase in population, economic growth, tourist influx, and modernization. Shimla and its surrounding areas need to absorb this population growth and adapt to modernity without losing the sense of place. A strong sense of nostalgia was revealed among residents as well as returning tourists about preserving and promoting the British heritage.

This study examined current development practices of Shimla in terms of resident and tourist perception. It concluded that there is no significant difference between both groups; they gave equal importance to the quality of the environment. This type of development is not unique to Shimla alone. Similar issues can be observed in surrounding communities of Kausauli, Solan, Mandi, Kulu and Manali in Himachal Pradesh as well as other Himalayan hill towns like Mussoorie and Nainital in Uttarakhand, and Darjeeling in West Bengal. The study recommends that cities which draw tourists should be developed in the context of their history, architectural heritage, and landscape character. This research highlights the need to maintain and enhance the visual character of townscapes. Visually attractive cities not only bring civic pride to its residents but also attract tourists and increase economic activity.

For Shimla and other hill towns to continue as viable, attractive, and promising tourist destinations, planners must pay attention to the preservation and protection of their natural and cultural landscapes. They must also ensure that new development is based on the principles of sustainable development and not consumption of the natural environment.

References

Allen, R.L., P.T. Long, R. Perdue, and S. Kieselbach. 1988. The impact of tourism development on residents' perceptions of community life. *Journal of Travel Research* 27: 16–21.

Biswas, S. 2006. India pushes graveyard tourism. *BBC News- Asia*.http://news.bbc.co.uk/2/hi/south_asia/6209517.stm. December 27.

Blake, P. 1964. *Gods own junkyard*. New York: Holt, Rinehart and Winston.

Burke, B. 1976. *Townscapes*. London: Penguin.

Butler, R.W. 1980. The concept of tourism area cycle of evolution: Implication for management of resources. *The Canadian Geographer* 24(1): 5–12.

Caneday, L., and J. Zeiger. 1991. The social economic, and environmental costs of tourism to gaming community as perceived by its residents. *Journal of Travel Research* 30: 45–48.

Cullen, G. 1961. *The concise townscape*. London: Architectural Press.

Dhaleta, S. 2013. Tourist season, traffic snarls peak in Shimla. *The Times of India*.http://timesofindia.indiatimes.com/city/chandigarh/Tourist-season-traffic-snarls-peak-in-Shimla/articleshow/20388115.cms. June 2.

Doxey, G. V. 1975. A causation theory of visitor-resident irritants, methodology and research inferences. In *Conference proceedings: Sixth annual conference of travel research association*, Vol. 1, 195–198. San Diego.

Gruen, V. 1964. *The heart of cities*. New York: Simon and Schuster.

Jurowski, C., M. Uysal, and D.R. Williams. 1997. A theoretical analysis of host community residents reactions to tourism. *Journal of Travel Research* 36: 3–11.

Jutla, R. 2000. Visual image of the city tourists versus residents' perception of Simla a hill station in northern India. *Tourism Geographies* 2: 404–420.

Kanwar, P. 1997. Can we afford Raj-type Simla? *The Tribune*, July 12.

Kavallins, I., and A. Pizam. 1994. The environmental impact of tourism – Whose responsibility is it anyway? The case study of Mykonos. *Journal of Travel Research* 36: 26–31.

Krippendorf, J. 1997. *The holidays maker: Understanding the impact of leisure and travel*. Trans. Vere Andrassy. Oxford: Heinemann.

Law, C. 1996. Introduction. In *Tourism in major cities*, ed. C. Law. London: International Thomson Business Press.

Lawson, R.W., J. William, T. Young, and J. Cossens. 1998. A comparison of residents' attitudes towards tourism in 10 New Zealand destinations. *Tourism Management* 19: 247–256.

Lohumi, R. 1999. From obscure hamlet to a renowned hill resort. *The Tribune*, May 5.

Lynch, K. 1996. *Image of the city*. Cambridge, MA: The MIT Press.

Middleton, V., and R. Hawkins. 1998. *Sustainable tourism: A marketing perspective*. Oxford: Butterworth-Heinemann.

Page, S. 1995. *Urban tourism*. London: Routledge.

Price, E. 1964. Velerbo: Landscape of an Italian city. *Annals of the Association of American Geographers* 54: 242–275.

Ryan, S., A. Scotland, and D. Montgomery. 1998. Resident attitudes to tourism development – A comparative study between the Rangitikei New Zealand and Bakewell, United Kingdom. *Progress in Tourism and Hospitality Research* 4: 115–130.

Sethi, N. 2013. Uttarakhand disaster got magnified due to heavy pilgrim rush. *Times of India*. http://timesofindia.indiatimes.com/india/Uttarakhand-disaster-got-magnified-due-to-heavy-pilgrim-rush/articleshow/20692148.cms. June 21.

Sharma, S.P. 2000. Mindless growth. *The Tribune*, March 25.

Sharma, A. 2001. Hazard on the highway. *The Tribune*. http://www.tribuneindia.com/2001/20010811/windows/main1.htm. August 22.

Singh, T.V. 1985. The paradox of mountain tourism: Case reference from the Himalayas. *Industry and Environment* 9: 21–26.

Singh, T.V. 1989a. On developing Himalayan tourism ecology. In *Studies in Himalayan ecology*, ed. T.V. Singh and J. Kaur. New Delhi: Himalayan Books.

Singh, T.V. 1989b. *The Kulu Valley: Impact of tourism development in the mountain areas*. New Delhi: Himalayan Books.

Singh, T.V. 1991. The development of tourism in mountain environment: The problem of sustainability. *Tourism Recreation Research* 16: 3–12.

Singh, M. 2000. Queen of the hills or hills of concrete? *The Tribune*, August 1.

Singh, T.V., and J. Kaur. 1985. *Integrated mountain development*. New Delhi: Himalayan Books.

Smailes, A.E. 1955. *Some reflections on geographical description and analysis of townscapes*. London: Transactions of Institute of British Architects.

Tringano, G. 1984. Tourism and the environment: The club Mediterranean experience. *United Nations Environmental Program Industry and Environment* 7: 20–21.

Whyte, W. 1968. *The last landscape*. New York: Double Day.

Whyte, W. 1980. *The social life of small urban spaces*. New York/Washington, DC: Conservation Foundation.

Chapter 10
Community-Based Tourism and Development in the Periphery/Semi-periphery Interface: A Case Study from Viet Nam

Tuan-Anh Le, David Weaver, and Laura Lawton

Abstract This chapter investigates how local residents in community-based tourism (CBT) situations within the periphery/semi-periphery interface perceive the relationship between CBT and 'development'. Qualitative research methods using a grounded theory approach were applied to examine perceptions of 55 local residents living in three CBT case study sites in the hinterland of Sa Pa town, a popular tourist destination in north-western Viet Nam. The study shows that residents not only regard development as a complex and multi-dimensional construct but overwhelmingly recognize the positive role played by CBT, and identify the influencing factors and recommendations to optimize the contributions of CBT for relevant stakeholders.

Keywords Community-based tourism • Grounded theory • Periphery • Semi-periphery • Resident perceptions • Viet Nam

10.1 Introduction

Tourism, increasingly, is regarded as a particularly effective and viable tool for development in rural areas and emerging economies, abetted by an annual average growth in global overnight arrivals of 7 % since 1950 (UN 2010). In 2012, for the first time, over one billion international overnight arrivals were recorded, and 1.6 billion arrivals were expected by 2020. It is anticipated, further, that most of this growth will occur in the Asia-Pacific region (UNWTO 2013). Most tourism can be described as large-scale or 'mass tourism', but sustained criticism of the latter since the 1970s gave rise to interest in 'alternative tourism' (Weaver 1998). As an ideal type, this evokes small-scale, locally controlled, culturally authentic, and

T.-A. Le (✉)
Ha Noi University, Hanoi, Vietnam
e-mail: hrdtourism@gmail.com

D. Weaver • L. Lawton
Griffith University, Gold Coast, Australia

S.F. McCool, K. Bosak (eds.), *Reframing Sustainable Tourism*,
Environmental Challenges and Solutions 2, DOI 10.1007/978-94-017-7209-9_10

highly regulated tourism practices (Weaver and Lawton 2014). Community-based tourism (CBT) emerged during this era as a subset of alternative tourism and was regarded by advocates as synonymous with sustainable tourism because of its emphasis on empowering and benefitting local residents (Murphy 1985; Weaver 1998; Hatton 1999; Scheyvens 1999; Leksakundilok 2004; Beeton 2006). Given the opportunities it seemingly creates for diversifying the local economy and providing equitable income and other economic benefits, CBT has become widely adopted by development agencies as a preferred and allegedly more effective tool for achieving economic and social development, particularly in rural and underdeveloped areas where options for alternative economic development are constrained by geographical and economic factors.

Although the number of existing CBT initiatives is impressive (Zeppel 2006), it appears that most have not lived up to their alleged potential to sustain rural communities (Salafsky et al. 2001). A paucity of necessary tourism-related managerial and technical capacity and knowledge is one factor (Blackstock 2005; Kontogeorgopoulos 2005; Li 2006), while Okazaki (2008) contends that local residents do not adequately understand the parameters of effective participation. The assumption that local residents all work together harmoniously and equitably with shared power toward common objectives has also been challenged (Weaver 1998; Beeton 2006), and while interaction and communication with relevant 'outside' stakeholders (e.g. investors, developers, planners and managers from the outside community) is critical to the success of any CBT, the latter is usually characterized by dependency on external development and funding agencies which may not share common goals or motivations (Butcher 2007; Goodwin and Santilli 2009; Weaver 2010). In particular, there is increasing concern that CBT is overly reliant on Western 'experts' and development organizations who impose their own – often ideologically driven – assumptions about the nature of 'development' and the optimal means for achieving same. Butcher (2007) describes how non-governmental organizations (NGOs) often exercise dictatorial power over local communities so that their 'development' conforms to what is wanted by the NGOs rather than the residents. It is therefore questionable whether any real community empowerment exists in CBT (Butcher 2010), especially at the highest level of self-mobilization where residents participate by taking initiatives independently of external institutions (Pretty 1995).

Despite the empowerment rhetoric of CBT, far too little attention has been paid to local and non-Western perspectives and knowledge. There is presently no study which adequately explores how local communities actually perceive 'development', how CBT contributes to achieving that goal, and how CBT can actually be improved. At the same time, Western CBT discourses typically assume a 'pure' peripheral context dominated by cohesive community structures and traditional livelihoods. This neglects the fact that many CBT projects are found in the dynamic interface between the periphery and the rapidly expanding semi-periphery which represents the 'front lines' of contemporary development and offers a complex and transitional landscape of contrasting environmental, cultural, social and economic impulses.

To address the attendant gaps in the literature and enhance the sustainability of CBT, this chapter investigates how local residents in CBT situations within the periphery/semi-periphery interface perceive the relationship between CBT and 'development'. Qualitative research methods using a grounded theory approach were applied to examine perceptions of the performance of CBT in three communes in the hinterland of Sa Pa, a popular tourist destination in north-western Viet Nam. This country is relevant because of the government's encouragement of CBT as a vehicle for facilitating rural development and thereby reducing pronounced internal core-periphery disparities. Residents of the case study region, given the opportunity to voice their own opinions, were asked (a) how they perceived 'development', (b) how CBT contributed toward the latter, (c) what factors positively and negatively influenced this contribution, and (d) what could be done to increase this contribution. Such lines of inquiry, and especially (a) and (d), are entirely consistent with the community-focused paradigm of CBT and break new ground in our understanding of this phenomenon and its mobilisation as an effective vehicle for development.

10.1.1 Core, Periphery and Semi-periphery

Major development paradigms of the latter half of the twentieth century have been implicitly or explicitly informed by core-periphery dynamics. The terms 'centre' and 'periphery' were initially introduced by Prebisch (1950) who contended that "the global system was not a uniform marketplace with producers and suppliers freely making mutual beneficial contracts but was in fact divided into powerful central economies and relatively weak peripheral economies" (Preston 1996). 'Core' and 'Periphery' differentials can be distinguished through (a) economic indicators [e.g. Gross Domestic Product (GDP) per capita and percentage employed in non-primary sectors], (b) exchanges of goods and services, and (c) patterns of interaction (Galtung 1971; Wellhofer 1989). Typically, cores exhibit higher levels of economic development, higher material standards of living, more highly skilled workers, higher levels of urbanisation and larger capital investments per worker. As such, they are the centres of political power, cultural domination, and military superiority (Wallerstein 1974). Reflecting their dynamic character, 'core–periphery' structures occur not only at the state level, but also internally, with each state exhibiting its own 'core' and 'periphery' differentiations (Galtung 1971).

Post-1950 modernization discourses assumed that cores were incubators of innovation and economic growth from which developmental impulses gradually 'trickled down' to the periphery through contagious and hierarchical diffusion effects, leading eventually to a more fully articulated space-economy (Rostow 1959). Dependency theory, subsequently, inverted this dynamic by contending that certain advantaged areas became cores by actively 'underdeveloping' other areas and initiating unequal terms of exchange. The periphery, accordingly, is created and reinforced by the perpetuation of those exchanges (Amin 1974). CBT, as an affiliate of alternative tourism, appeared in effect as a mechanism for maintaining or reinstating the empowerment of local communities in the periphery in the face of these underdevelopment processes. Post-dependency paradigms such as sustainable development and globalization are less ideologically polarized but similarly invoke core and periphery as a fundamental dynamic in global and regional space economies.

Wallerstein (1984), in his dependency-inspired World Systems theory, expanded the 'core-periphery' concept to include the 'semi-periphery' as a transitional space, independent of state boundaries (Terlouw 2001), which depolarizes and mediates the core-periphery relationship. Semi-peripheries are partially industrialized, but with less sophisticated technology than in the core. In Dependency and World Systems discourses, it is exploited by the core but in turn exploits the periphery through trickle-up processes. However, in the Modernization perspective, semi-peripheries serve as avenues for disseminating the benefits of development to adjacent peripheries through trickle-down processes; as a result, those peripheries themselves eventually become part of the semi-periphery. This idea of 'avenue' can be taken literally in situations where the semi-periphery occurs as a highway, river or railway corridor which attracts adjacent development and expedites contact between core and periphery. In post-dependency discourses, the dynamic element is pervasive as globalization impulses extend the semi-periphery into the periphery along all fronts. Based on the literature, the core, periphery and semi-periphery can be differentiated as per Table 10.1.

Table 10.1 Core, periphery and semi-periphery

	Core	Semi-periphery	Periphery
Economics	High levels of vitality and a diverse economic base	Medium levels of vitality and an economic mix of activities with lax regulation and strong developmental pressures	Low levels of vitality and dependent on traditional industries
Population	Metropolitan in character. Rising population through in-migration with a relatively young age structure	Semi-urban and strong population pressures	More rural and remote – often with high scenic values. Population falling through out-migration, often with an ageing structure
Technology/ communication	Innovative, pioneering, and enjoys good information flow	Medium technology and information flow at average level	Reliant on limited and imported technologies and ideas, and suffers from poor information flows
Power/decision	Focus of major political, economic and social decisions	Subordinate to the core, but has resources to resist or co-operate	Remote from decision making leading to a sense of alienation and lack of power
Infrastructure	Good infrastructure and amenities	Basic infrastructure and amenities	Poor infrastructure and amenities
Evolution	Stable	Expanding	Contracting

Source: Modified from Botterill et al. (2000), Terlouw (2003) and Wanhill (1997)

However spatially construed, the identification of appropriate development metrics is a major concern for all development theorists, since, presumably, a primary concern of all is to determine the extent to which a particular country or region has achieved ‘development’. Since the 1950s, various indicators/metrics have been used to measure ‘development’, covering not only economic development but also socio-political parameters such as human well-being and quality of life. The evolution of these metrics is on-going, evolving from econometrics such as productivity and GDP (Rostow 1959); GNP (McGranahan 1970); GDP per capita, percentage employed in non-primary sectors and good values exchanged between core and periphery (Galtung 1971; Wellhofer 1989), to more sophisticated, diverse and complex constructs incorporating both economic and non-economic factors (Myrdal 1974). These include the Human Development Index (HDI) (UNDP 1990), the Inequality–Adjusted Human Development Index (IAHDI) (Hicks 1997), and globally accepted Millennium Development Goals (MDGs) (UN 2000), all of which reflect holistic quality of life aspirations ranging from improving poverty rates, promoting gender equality, and providing universal primary education to ensuring environmental sustainability. It is noted that development and its measurement is mainly assessed by international organizations such as the UNDP, World Bank, international NGOs and other outside experts who assume their own expertise and do not adequately take into account the opinions of local residents.

10.1.2 Context of Viet Nam

Viet Nam, with almost 90 million inhabitants, displays complex post-World War Two development patterns, with the north initially being influenced by Marxism and the south by USA supported modernization. After reunification in 1976, Viet Nam, categorized by Wallerstein (1974) as a semi-peripheral state, continued to emphasize central planning and economic subsidies. This period ended with the introduction of the “Doi moi” (open-door) policy in 1986, which contributed to Viet Nam’s successful transition toward a ‘socialist-oriented market economy’ (Irvin 1995) that amalgamates Marxist, modernization, globalization, sustainability and dependency impulses. Increased integration into the global economy is indicated by membership in peak forums such as ASEAN, APEC and the WTO. Largely as a consequence of this integration, Viet Nam’s development metrics are rapidly improving (Table 10.2). Notably, the percentage of the population below the national poverty line of US$1 per day has declined from around 75 % in 1990 to under 11.3 % in 2012 (Centre for International Economics 2002, p. 2; Central Intelligence Agency 2014). In 2010, Viet Nam entered middle income country status, a step toward the official government goal of becoming a modernized and industrialized country by the year 2020 comparable to its regional neighbors Malaysia and Thailand.

In 2000, Viet Nam committed to the implementation of the Millennium Declaration and integrated the MDGs into the National Socio-Economic Development Strategy (2000–2010). The MDGs have been recognized through the Comprehensive

Table 10.2 Overview of Viet Nam development indicators

	2008	2009	2010	2011	2012	2012 global ranking
GDP (purchasing power parity – PPP) (US$ billion)		*256.9*	*293.1*	*310.4*	*325.9*	*38/229*
GDP per capita (PPP)		*2.900*	*3.300*	*3.500*	*3.600*	*167/229*
GDP (official exchange rate) (US$ billion)				*123.6*	*138.1*	*57/193*
Poverty (% pop. living below national poverty line of $1 per day)	*13.4*				*11.3*	
GNI Atlas method (current US$ million)	*78.439*	*88.537*	*101.089*	*111.125*		
GNI per capita (2005 PPP$)	*2.740*	*2.850*	*2.757*	*2.859*	*2.970*	
Human Development Index (HDI)	*0.597*	*0.601*	*0.611*	*0.614*	*0.617*	*127/187*
Inequality – adjusted HDI			*0.478*	*0.510*	*0.531*	*70/132*
Gender Inequality Index				*0.305*	*0.299*	*48*
Life expectancy at birth, total (years)	*74*	*74.4*	*74.9*	*75.2*	*75.4*	*94/223*
Total fertility rate (children born/woman)			*1.93*		*1.89*	*135/224*
Mortality rate, infant (per 1,000 live births)		*69*			*19.61*	*95/224*
Literacy rate (% ages 15 and over can read and write)	*92.5*			*93.4*		

Sources: Complied from Central Intelligence Agency (2012, 2014); General Statistic Office of Viet Nam (2010); UNDP (2012); World Bank (2012)

Poverty Reduction and Growth Strategy (2002), which defines the adopted Viet Nam Development Goals (VDGs). These take into account the specific development features of Viet Nam and include additional goals of reducing vulnerability, improving governance for poverty reduction, reducing ethnic inequality and ensuring pro-poor infrastructure. Abetted by a strong rhetorical commitment to sustainable development (Nguyen 2010), Viet Nam was expected to achieve most of the MDGs by 2015. However, there exist challenges in particular on disparities between the Kinh majority and ethnic minority groups, and the lack of drinking water and sanitation (UNDP 2010; World Bank 2010). With regard to these and all other development indicators, Viet Nam displays notable core-periphery differentiations. Based on General Statistic Office data for the years 2008–2012 concerning area, GDP per capita, population and accessibility (measured by the presence of road transportation infrastructure), tentative core–periphery and semi-periphery areas/corridors in Viet Nam have been mapped by the authors (Fig. 10.1).

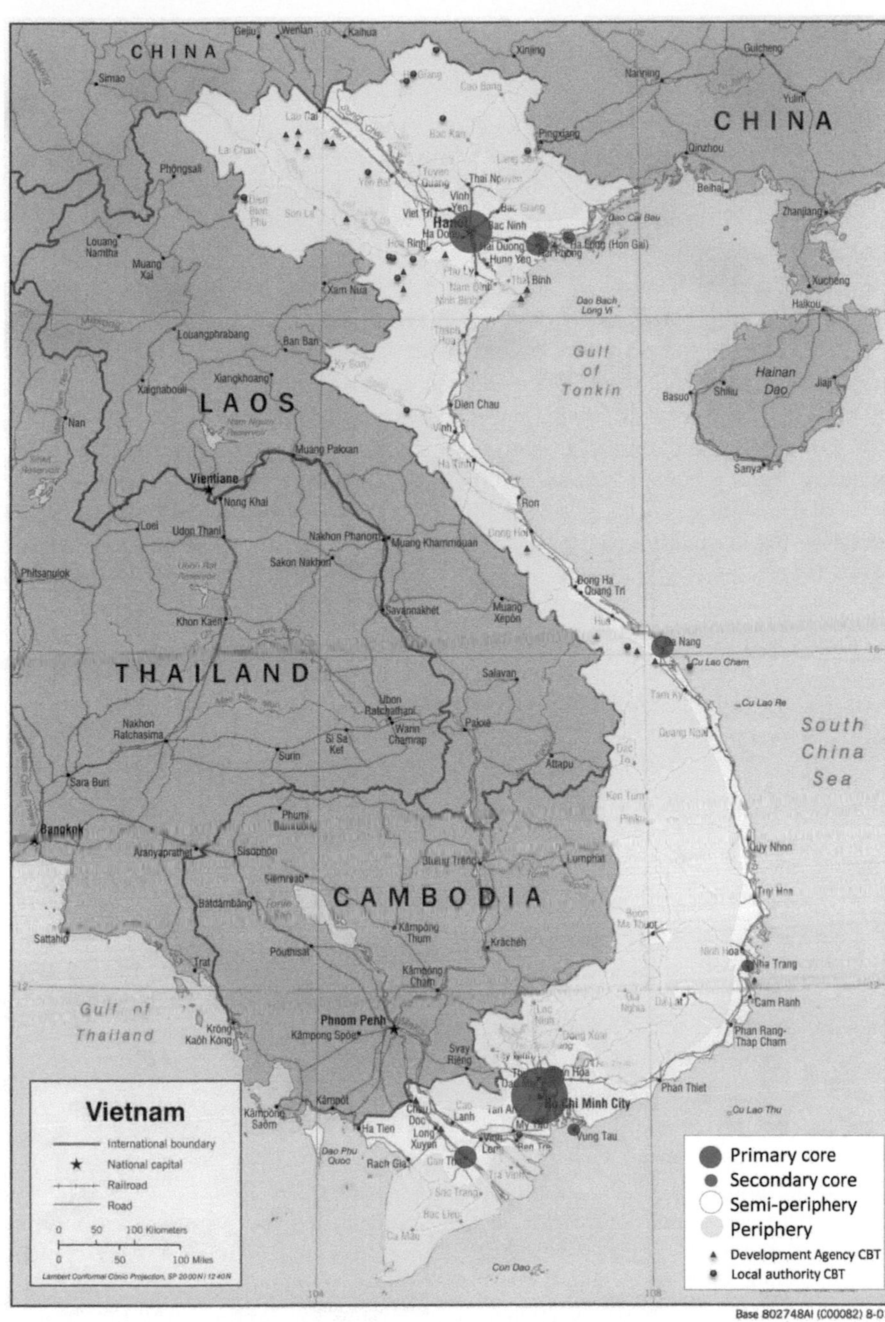

Fig. 10.1 Core, semi-periphery and peripheral areas of Viet Nam

Hanoi and Ho Chi Minh City are the two primary city-region cores, characterized by high levels of economic vitality and good infrastructure. Together they occupy only 1.6 % of Viet Nam but account for 16 % of the population and over 30 % of GDP (General Statistic Office of Viet Nam 2010). There are also several secondary cores, each functioning as a second-tier urban hub within its respective region, i.e. Hai Phong and Ha Long cities in the North, Da Nang and Nha Trang cities in the Central and Ba Ria-Vung Tau, Binh Duong and Can Tho in the South. These secondary hubs collectively account for 7 % of Viet Nam's land area, 11.3 % of population and 31 % of GDP. Together the primary and secondary represent just 9 % of land area but 27 % of population (53 % urban population) and 62 % of GDP. In contrast, the periphery encompasses remote mountainous and borderland areas where infrastructure is poor and the economy is dependent on traditional industries. These areas account for 87 % of land area and 67.5 % of population (19 % of urban population), but just 28 % of GDP. Ethnic minorities comprise the majority of the periphery's population. The semi-periphery includes the remaining 4 % of land area, 5.5 % of population (28 % of urban population) and 10 % of GDP. These mainly rural areas are characterized by a high level of sedentary agricultural activity and relatively good connectivity with local service centres and transportation networks.

Since the introduction of the open door policy in the late 1980s, Vietnamese tourism has diversified its markets and activities, and, in tandem with the broader Asia-Pacific region, achieved a remarkable average annual increase in international tourist arrivals of 17.8 % during the period 1990–2011. It received only 250,000 inbound tourists in 1990, but more than six million in 2011 with US$6.2 billion in receipts. Domestic tourism concurrently increased by 20.3 % per year, from one million in 1990 to 30 million in 2011, reflecting the rapid growth of the middle class in the core. Tourism in 2012 accounted for 5.8 % of national GDP (Viet Nam National Administration of Tourism 2012; World Travel and Tourism Council 2012). The Tourism Development Strategy for the period 2011–2020 envisages "tourism as an important economic sector, generating foreign income, employment and contributing to transformation of economic structure" by 2030. An annual increase from 10 % to 15 % in inbound tourists and a doubling of the current tourism contribution up to 8 % of GDP by the year 2020 has been targeted (Government of Viet Nam 2011).

The earliest Vietnamese CBT initiatives in 1985 all involved periphery locations and were focused on Eastern European markets SNV (The Netherlands Development Organisation 2007). With the subsequent open door policy, CBT was extended to additional peripheral locations and opened to Western and domestic tourists, aided by the concerted assistance of international NGOs and development agencies such as SNV (Netherlands Development Organization), AECI (Spanish Development Agency), IUCN (International Union for Conservation of Nature), FFI (Fauna and Flora International), and WWF (World Wildlife Fund/World Wide Fund for Nature). Currently there are estimated to be 35 CBT projects/initiatives in Viet Nam. Geographically, one-third of these are located in the transitional zone between periphery and semi-periphery and the remainder are in 'pure' peripheral areas. Most involve ethnic minority groups. As noted earlier, the semi-periphery is increasingly implicated in CBT but has not yet been positioned or examined as a context distinct from the 'pure' peripheral settings that dominate the literature. Moreover, the issue of constrained community empowerment that derives from the excessive influence of NGOs and development agencies is exacerbated by Viet Nam's tradition of central planning, which privileges government as the primary source of knowledge and decision-making authority.

10.2 Methods

Qualitative research design was used to provide an in-depth, descriptive and interpretive analysis of local community perceptions. More specifically, objectivist grounded theory (Strauss and Corbin 1990; Strauss and Corbin 1998; Corbin and Strauss 2008) was applied to investigate three case study communes in the transitional zone between the periphery and semi-periphery in north-western Viet Nam's Sa Pa region, which possesses a concentration of CBT initiatives. Such methods allow the researcher to investigate personal experiences (Strauss and Corbin 1990), seek experiential lived community 'knowledge', and study "things in their natural settings, attempting to make sense of or interpret phenomena in terms of meanings people bring to them" (Denzin and Lincoln 2005). Accordingly, the inquirer generates a general explanation (a theory) of a process, action, or interaction shaped by the views of a large number of participants (Strauss and Corbin 1998; Corbin and Strauss 2008). Grounded theory has been widely used in tourism research (Connell and Lowe 1997; Hardy 2005; Weaver and Lawton 2008; Lawton and Weaver 2009; Castellanos-Verdugo et al. 2010), and is appropriate here as the study seeks ultimately to develop an optimal CBT "grounded" in data from local residents at the selected sites, who might help explain practice or provide a framework for further research (Creswell 2007). Semi-structured in-depth interviews were conducted with 55 local residents, the number at each site being dictated by researcher assessments of theoretical saturation (Weaver and Lawton 2008; Lawton and Weaver 2009; Bryman and Bell 2011; Bryman 2012). Residents were interviewed individually so as not to be unduly influenced by the presence of other villagers, and the researcher, beyond asking a consistent series of basic questions, took care not to influence or bias the responses.

The case study approach is used extensively in tourism research (Beeton 2004; Xiao and Smith 2006; Kibicho 2008). According to Yin (2009), the case study is an empirical inquiry that investigates a contemporary phenomenon in depth and within its real-life context, especially when the boundaries between phenomenon and context are not clearly evident. It is a preferred method when (a) "How" or "Why" questions are being posted, (b) the investigator has little control over events, and (c) the focus is on the contemporary phenomenon within a real-life context. All three factors, notably, concurrently justify the grounded theory approach. As analytic conclusions independently arising from more than a single case will be more powerful, and especially if each case offers some contrasting results, multiple-case design is selected to supplement the grounded theory approach as the evidence is more compelling and robust, and there is a high possibility of replication (Stake 1995, 2006; Yin 2009).

10.3 Sa Pa as a Regional Focus

Sa Pa, with 52,000 residents, is located along the mountainous interface between the periphery and a semi-peripheral corridor in Lao Cai province in north-western Viet Nam. Besides the Vietnamese people (17.9 %) there are five main ethnic minority groups in Sa Pa, i.e., Hmong (52 %), Dao (23 %), Tay (4.7 %), Giay (1.4 %) Xa Pho, and a small number of other ethnic minorities. The region is noted for outstanding natural and cultural resources, and is accessible from Hanoi by overnight train or bus. Most international travellers use the town of Sa Pa as a gateway for accessing nearby CBT villages, which provide accommodation and exposure to ethnic minority cultures during a 1 or 2 day trek. They may also participate in voluntary community development work such as renovation of a commune school or clinic, teaching foreign languages and planting trees. Domestic tourists prefer to visit communes during the day or walking around the town (Fig. 10.2).

Sa Pa was a tourist destination during the French colonial period but was destroyed and ignored because of the war with the USA that ended in 1975 and conflict with China from 1979 to 1990. Sa Pa was only revitalized in the early 1990s and has since experienced rapid tourism development. Between 2000 and 2013, tourist arrivals have increased from 49,322 to 610,000, 20 % of whom were international (Sa Pa District People Committee 2013) (Fig. 10.3).

With regard to visiting villages, day visitors dominate all sites (Table 10.3). To ensure that the development is sustainable and beneficial to local communities rather than only to tourism enterprises or the relatively developed semi-peripheral town of Sa Pa, SNV together with IUCN launched a 3-year project (2001–2003) titled 'Support Sustainable Tourism in Sa Pa District'. Under this project, CBT and its attached products of responsible trekking routes and homestays in minority communes was introduced to the hinterland of Sa Pa (SNV 2007). Currently, five communes offer 99 homestay accommodation services, mostly in Ta Van (48.5 %) and Ban Ho (31.3 %).

The Sa Pa region was selected as a regional focus because of its well-defined periphery/semi-periphery interface and its inclusion of CBT within a rapidly growing regional tourism sector. Three nearby villages – Ban Ho, Ta Van, and Ta Phin – were selected as case studies. Here, in contrast to the modern and developed town of Sa Pa, where ethnic Vietnamese dominate the services-oriented economy, numerically dominant ethnic minority peoples are becoming increasingly engaged with tourism as a supplement to traditional livelihoods focused around farming, forestry and handicrafts. In each of the three villages, there is: (a) a formal CBT management board in place but (b) a different dominant ethnic minority group, (c) a

Fig. 10.2 Map of Sa Pa and CBT case studies

different distance from Sa Pa town, (d) a different population size, and (e) a different apparent stage of the destination life cycle (Fig. 10.4).

10.3.1 Ban Ho Commune

Located 24 km southeast of Sa Pa, Ban Ho has 510 household with 2709 residents (15.26 % below the poverty line of US$1 per day in 2009) with ethnic minorities accounting for 80 %. Tourism's introduction in 1997 is attributable to the natural beauty of a local waterfall and hot water streams as well as brocade products, traditional cultural performances, and hospitable people. Currently, 31 traditional

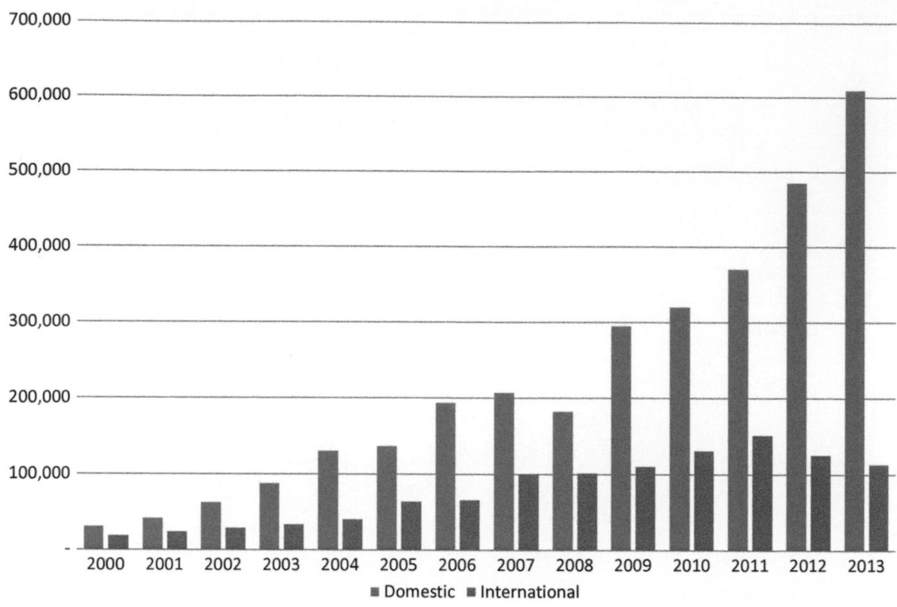

Fig. 10.3 Tourist visiting Sa Pa (2003–2013)

Table 10.3 Visitor arrivals at CBT sites in Sa Pa

No.	CBT site	Year 2010		Year 2011		Year 2012	
		Day visitor	Stay-over	Day visitor	Stay-over	Day visitor	Stay-over
1	**Ta Van**	102,532	12,321	131,882	18,672	136,800	16,200
2	**Ta Phin**	36,856	236	39,317	533	47,800	854
3	**Ban Ho**	6,541	2,530	10,231	3,380	17,200	2,418
4	**Thanh Phu**	1,952	652	2,883	1,176	1,650	1,213
5	**Sin Chai**	860	103	2,589	213	3,200	251
Total		**148,741**	**15,842**	**186,902**	**23,974**	**206,650**	**20,936**

Source: Sa Pa People Committee Report (2013)

wooden houses-on-stilts, belonging to the Tay minority, offer tourists homestay services, soft drinks and food. Over 10,000 day visitors and 5461 stayover tourists visited Ban Ho in 2008. However, this number declined 50 % to 5000 day visitors and 2991 stayover in 2009 due to environmental damage caused by the construction of two local hydro power stations. This induced the decline stage of the destination life cycle (Fig. 10.4) as the balance between economic development and tourism was not maintained. The situation has improved gradually since 2010 with the completion of one hydro power station (Su Pan 2). Despite positive indicators that visitors are returning, observation shows that most of them are day-only excursionists, with over 17,000 of the latter visiting in 2012. The local

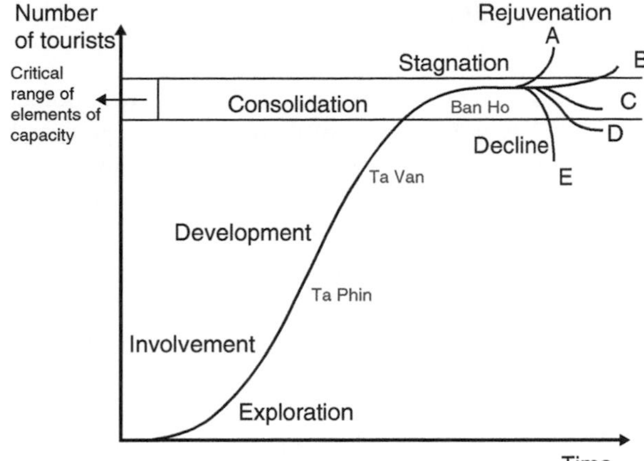

Fig. 10.4 Stages of CBT sites on the destination life cycle. Adopted from Butler (1980)

CBT management board started its operation since 2008. In the first year of its operation 2008/2009, the board registered about 5000 stayover visitors, generating VND350,000,000 in revenue (US$17,500: US$1 is equal to VND20,000). This revenue was used to pay an allowance of VND500,000 per person for five working staff, and to establish a community development fund.

10.3.2 Ta Van Commune

Located in Muong Hoa valley, about 8 km in the South of Sa Pa town, Ta Van is famous for its terrace fields, which were voted as one of the seven most attractive terraced fields in the world by the readers of Travel and Leisure Magazine (Hoang 2009). There are 689 Giay ethnic households (3791 people) of which 37.28 % are below the poverty line. Apart from beautiful scenery, Ta Van attracts tourists with its peaceful lifestyle, skilfully carved silver products, local customs, and ethnic food. In addition, the Giay people live both in stilt-houses and land houses. Because of its proximity to Sa Pa town, Ta Van has received an increasing number of tourists, reaching 136,800 in 2012. Stayovers have increased from 6000 in 2007 to 16,200 in 2012, which positions the commune at the late development stage of the Butler cycle. The CBT management board started operation in 2008, when it registered 10,306 stayovers. The homestay revenue is more than VND500,000,000 (US$25,000) in 2008/2009. A community development fund has been established and five staff were supported with monthly allowance of VND500,000 (US$25) each.

10.3.3 Ta Phin Commune

Ta Phin is 12 km northeast of Sa Pa town. It is home to 566 red Dao households with 2997 people (41.3 % of them below the poverty line) and well-known for its monastery, a cave with attractive stalactites, and hand-made brocade products. CBT was introduced to Ta Phin through a project implemented jointly by Hanoi Open University (Viet Nam), and Capilano College and North Island College (Canada) with financial support from CIDA (Canadian International Development Agency) from 2005 to 2008 and PATA foundation funding from 2010 to 2014. Ta Phin is now one of the main tourist attractions of Sa Pa region, receiving about 50,000 visitors in 2012, including 854 stayovers. Five homestays offer accommodation and food services. Visitation trends suggest that the commune situates in the mid-development stage of the destination life cycle. The CBT management has been established since 2010, when 1500 stayovers were registered. The tourism revenue from of homestay was VND140,000,000 (US$7000) in 2010/2011, which is used to support five staff. In addition to a community development fund, the Board set up an embroidery club and a display centre to showcase and sell local products to visitors for extra income.

10.4 Results

10.4.1 Local Residents' Perceptions of 'Development'

The most salient finding was the primacy of economic considerations, with income growth nominated overwhelmingly as the main indicator of 'development'. However, about two-thirds of interviewed residents subsequently qualified this econometric perspective with socio-cultural and environmental criteria, revealing 'development' as a complex and multi-dimensional construct. For residents of Ban Ho, environmental recovery was regarded as a very important aspect of development, while the environmental dimension in Ta Phin was alternatively conceived in terms of good hygiene. Residents in Ta Van and Ta Phin cited opportunities for interacting with the outside world as an important socio-cultural parameter. For residents of Ban Ho and Ta Phin, retention of traditional craft skills was cited as a secondary indicator of development.

10.4.2 Role of CBT in Facilitating 'Development'

Residents in all three villages not only regard income growth as the primary indicator of development, but overwhelmingly recognize the role played by CBT in improving their revenue stream and, concurrently, generating employment –

tourism, accordingly, is equated with development. As with development, complexity is evident in tourism with the invoking of socio-cultural and environmental facets by over half of respondents. CBT, through contact with tourists, is credited with helping to attain greater connectivity with the outside world, creating more opportunities for education, and with stimulating demand for a better physical environment (whether in terms of biodiversity or hygiene) and for traditional handicrafts. 'Development', while therefore strongly associated with the introduction of tourism, was also seen as entailing various costs, although these did not outweigh the perceived benefits. Economic costs included inflation and increased dependency on tourism, while socio-cultural costs included aggressive selling tactics by some vendors and conflict within the community. Interestingly, none of the residents cited any negative environmental impacts from CBT.

10.4.3 Factors Influencing CBT

Factors that influence the performance of CBT from the villagers' perspectives are divided into socio-cultural, economic, environmental, infrastructural, organizational/political, and spatial categories. The major finding shows that the majority of villagers perceived traditional cultures and beautiful scenery as the main facilitators of CBT because of their attractiveness to visitors. Also frequently cited were good infrastructure and proximity to Sa Pa. The most strongly expressed negative influence was the modification of traditional culture, while aggressive vendors, land speculation and over-dependence on tourism were also widely cited. Other problems included unattended trekking routes, lack of leadership, and limited length of visitors' stay. Differences are apparent among the villages. For example, while 'cultural modification' was universally important, 'house architecture' was perceived as the important specific factor in Ban Ho, while 'traditional costumes' were mentioned commonly in the other villages. Only in Ban Ho was environmental degradation perceived as a major impediment. In addition, both children and adults were perceived as aggressive vendors in Ta Phin while this phenomenon referred only to children (though not as a critical theme) in the other two villages.

10.4.4 Local Recommendations to Optimize the Contribution of CBT

The recommendations are diverse and grouped into socio-cultural, economic, environmental, infrastructural, product development and organizational/political categories. The major finding shows that the recommendations mainly came from a small number of villagers who are either senior or business- minded villagers [elite] directly involved in tourism businesses in the attendant villages. Though there are some very important similarities, the proposed recommendations are specifically relevant to addressing the identified negative factors at each village. Accordingly, special attention was called to preserve or revitalize traditional crafts, to promote and sell these in a more formalized manner, to raise prices in order to generate more income, and to restructure or establish the tourism management board.

10.5 Discussion and Conclusion

Contextually, there are tourism attributes in the three case study villages that implicate distinctive dynamics of the semi-periphery and influence subsequent resident perceptions. Paramount among these is the very high levels of visitation relative to resident populations, giving rise to the formation of 'hyperdestinations' (Weaver 2006). High tourist-to-resident ratios owe to the proximity of the villages to the regional gateway of Sa Pa, with distance-decay relationships being apparent to the extent that the closest of the villages, Ta Van, displays the highest such yearly-denominated ratio (51:1). Similarly, the great majority of visitors in each case consist of day-only excursionists, as in the rural–urban fringes of advanced economies which have hitherto dominated the hyperdestination literature. The CBT literature, in contrast, is strongly invested in the primacy of overnight accommodation as a critical source of revenue and dominant type of tourism-related infrastructure, catering to a relatively small number of visitors who tend to stay for at least several days (Hatton 1999; Kontogeorgopoulos 2005; Beeton 2006). Clearly, proximity to a major semi-peripheral gateway distorts and subverts such assumptions.

This proximity, and the transitional dynamics it implies, also reveals resident responses conflicted between the dictates of the periphery (tradition) and the semi-periphery, which functions as a conduit to and from the core (modernity). Regarding perceived definitions of 'development', the emphasis on income growth suggests a high degree of community integration into the conventional market economy that pervades the adjacent semi-periphery, and concomitant awareness of the importance of income as a means of obtaining education and other products indicative of improved quality of life. Sa Pa, in essence, is an increasingly important facet of the village space-economy through its role as an accessible source of retail goods and services. Concurrently, it is the primary source of the tourists who provide a growing proportion of the income that represents tangible development to these residents. Villagers make an explicit link between tourism and development because the former creates jobs and income, and provides an incentive for improving the quality of the environment and maintaining traditional handicrafts, which in turn reinforce tourism.

Residents, however, also recognize the costs of tourism, though as per social exchange theory (Ap 1992; Pfister and Wang 2008; Weaver and Lawton 2013), these are not regarded as exceeding the perceived benefits. Complaints about cultural modification, interestingly, display the ambivalence that is implicit in modernization, in that there is appreciation for material and financial gain but concern over the loss of tradition. It is ironic in this regard that residents acknowledge both this actual loss as well as the concurrent potential of tourism to *preserve* these traditions due to tourist demand. This pertains similarly to the appreciation of new tourism revenue but concern over increased dependency on this revenue. With regard to the town of Sa Pa, most residents explicitly cite proximity as a benefit and development facilitator because of the generation of large visitation numbers, but some also cite this proximity as a negative due to the dissuasive effect this has on stayover visitors, and the access this provides to external competitors – accounting perhaps for some of the aggressive vendors. Also embodying this sense of conflict between tradition and modernity is the evocation of beautiful scenery and traditional culture as the foundation assets for successful tourism. Both are profoundly connected with a relatively undisturbed periphery, yet both are potentially threatened by the developmental impulses of modernization, as demonstrated by the effects of dam construction in the vicinity of Ban Ho.

Much of the CBT literature laments the loss of tradition and the incursion of modernization impulses that at best are deemed by so-called experts to be inappropriate and ill-considered. The apparent tensions between tradition and modernity revealed in this periphery/semi-periphery interface certainly also exist in 'pure' periphery contexts, but seem to be exaggerated and exacerbated by proximity to regional urban gateways and transportation corridors. In theory, the resultant creation of hyperdestinations within a short period of time should be generating an array of economic, socio-cultural and environmental problems in tandem with the positioning of these three villages in stages of the destination life cycle that indicate the breeching of local carrying capacities (Butler 1980). Yet, contrary to expectations, these villagers appear to be very happy with their local tourism activity and welcome interaction with tourists. This however does not render them oblivious to the attendant problems, which display some variation according to the unique circumstances of each village.

Without subverting their own voices, it is ventured by the authors that villagers make sophisticated if implicit assessments of costs and benefits from tourism, and have determined in the main that the latter outweigh the former. Optimization and adjustment characterise these assessments, as for example in the desire to increase income and education while concurrently preserving traditional culture and protecting the natural environment. In so doing, tradition and modernity are regarded not as irreconcilable, but as complementary if seemingly contradictory impulses that together can contribute to enhanced quality of life. A similar dialectical balancing is evident in Kim Yujeong Literary Village in rural South Korea, where tourists outnumber locals by a yearly-denominated factor approaching 1500 to 1, yet residents are strongly supportive of the tourism industry and appear well adjusted (Lee and Weaver 2014). In Viet Nam and elsewhere, being on the periphery/semi-periphery interface may therefore be seen as more opportunity than threat with regard to the potential of CBT to improve the quality of life for residents.

10.6 Lessons Learnt

The first lesson learnt is respect for the local voice. The empowerment of local residents is crucial for the sustainable development of any CBT initiative. The residents are not only the most affected beneficiary but also a key stakeholder in ensuring the success of CBT. Moreover, notwithstanding widespread poverty and illiteracy, their ability to derive maximum benefits from tourism indicates a high level of sophistication and canniness. The universal cooperation of the interviewed residents, moreover, indicates that these stakeholders are enthusiastic about sharing their perceptions if appropriate relationships and trust are cultivated over time between the researcher and the researched. Additionally, the study showed demographic variation among communities thereby reflecting widespread local participation. The voice of elites who play a dual role as a knowledgeable resident within a community and an officer working in local authorities should be considered, but it must also be augmented by the voice of "ordinary" residents who also deserve to be heard.

The second lesson learnt is the importance of spatial context. The dynamic characteristics of periphery/semi-periphery interface areas, where tradition meets modernity in complex, dynamic and diverse ways, should be acknowledged when implementing any tourism/CBT activities. The findings of the study propose that there is potential to accommodate modernity in order to maximize tourist experiences and generate income while preserving the traditional culture and pristine environment – the primary resources to attract tourists.

The final lesson learnt for tourism planners, managers, and policy makers is that the application of the grassroots approach is essential in a country like Viet Nam where central planning and authoritarian leadership still prevail. The dual (dialectical) approach of the recommended grassroots and conventional top-down approaches could be negotiated to ensure sustainable development outcomes optimal for both the local residents and the good of the country as a whole. Tourism planning and management must take into account the voice of local resident.

References

Amin, S. 1974. *Accumulation on a world scale: A critique of the theory of underdevelopment*. New York: Monthly Review Press.

Ap, J. 1992. Residents' perceptions on tourism impacts. *Annals of Tourism Research* 19(4): 665–690.

Beeton, S. 2004. The case study in tourism research: A multi-method case study approach. In *Tourism research methods: Integrating theory with practice*, ed. B.W. Ritchie, P. Burns, and C. Palmer, 37–61. Cambridge, MA: CAB International.

Beeton, S. 2006. *Community development through tourism*. Collingwood: Landlinks Press.

Blackstock, K. 2005. A critical look at community based tourism. *Community Development Journal* 40(1): 39–49.

Botterill, D., R. Owen, L. Emanuel, N. Foster, T. Gale, C. Nelson, and M. Selby. 2000. Perceptions from the periphery: The experience of Wales. In *Tourism in peripheral areas*, ed. F. Brown and D. Hall, 7–38. Clevedon: Channel View.

Bryman, A. 2012. *Social research methods*. Oxford: Oxford University Press.

Bryman, A., and E. Bell. 2011. *Business research methods*. Oxford, UK: Oxford University Press.

Butcher, J. 2007. *Ecotourism, NGOs and development: A critical analysis*. London: Routledge.

Butcher, J. 2010. The mantra of 'community participation' in context. *Tourism Recreation Research* 35(2): 201–205.

Butler, R.W. 1980. The concept of a tourist area cycle of evolution: Implications for management of resources. *Canadian Geographer* 24: 5–12.

Castellanos-Verdugo, M., F.J. Caro-González, et al. 2010. An application of grounded theory to cultural tourism research: Resident attitudes to tourism activity in Santiponce. In *Cultural tourism research methods*, 115–128. Wallingford: CAB International.

Central Intelligence Agency. 2012. The world factbook. https://www.cia.gov/library/publications/the-world-factbook/geos/vm.html. Retrieved 8 Sept 2012.

Central Intelligence Agency. 2014. *The world factbook*. https://www.cia.gov/library/publications/the-world-factbook/geos/vm.html. Retrieved 8 Sept 2014.

Centre for International Economics. 2002. *Viet Nam poverty analysis prepared for Australian agencies for international development*, 118. Canberra: Centre for International Economics.

Connell, J., and A. Lowe. 1997. Generating grounded theory from qualitative data: The application of inductive methods in tourism and hospitality management research. *Progress in Tourism and Hospitality Research* 3(2): 165–173.

Corbin, J.M., and A.L. Strauss. 2008. *Basics of qualitative research: Techniques and procedures for developing grounded theory*. Los Angeles: Sage.

Creswell, J.W. 2007. *Qualitative inquiry and research design: Choosing among five approaches*. Thousand Oaks: Sage.

Denzin, N.K., and Y.S. Lincoln. 2005. *The sage handbook of qualitative research*. Thousand Oaks: Sage.

Galtung, J. 1971. A structural theory of imperialism. *Journal of Peace Research* 8(2): 81–117.

General Statistic Office of Viet Nam. 2010. *Statistical year book of Viet Nam [in Vietnamese]*. Ha Noi: Vietnam Statistics Press.

Goodwin, H., and R. Santilli. 2009. *Community-based tourism: A success?* ICRT occasional paper 11, 39. University of Greenwich, London, UK.

Government of Viet Nam. 2011. *Strategy on Viet Nam's tourism development until 2020, vision to 2030*. Ha Noi, Viet Nam: Ministry of Culture, Sports and Tourism.

Hardy, A. 2005. Using grounded theory to explore stakeholder perceptions of tourism. *Journal of Tourism and Cultural Change* 3(2): 108–133.

Hatton, M.J. 1999. *Community-based tourism in the Asia-Pacific*. Toronto: Humber College, School of Media Studies.

Hicks, D.A. 1997. The inequality-adjusted human development index: A constructive proposal. *World Development* 25(8): 1283–1298.

Hoang, K. 2009. *Sa Pa Viet Nam: A natural mosaic*. http://travelnews.activetravelshop.com/2009/10/sapa-vietnam-natural-mosaic.html. Retrieved 16 Sept 2010.

Irvin, G. 1995. Vietnam: Assessing the achievements of *Doi Moi*. *Journal of Development Studies* 31: 725–750.

Kibicho, W. 2008. Community-based tourism: A factor- cluster segmentation approach. *Journal of Sustainable Tourism* 16(2): 211–231.

Kontogeorgopoulos, N. 2005. Community-based ecotourism in Phuket and Ao Phangnga, Thailand: Partial victories and bittersweet remedies. *Journal of Sustainable Tourism* 13(1): 4–23.

Lawton, L.J., and D.B. Weaver. 2009. Travel agency threats and opportunities: The perspective of successful owners. *International Journal of Hospitality and Tourism Administration* 10(1): 68–92.

Lee, Y., and D. Weaver. 2014. The tourism area life cycle in Kim Yujeong literary village, Korea. *Asia Pacific Journal of Tourism Research* 19(2): 181–198.

Leksakundilok, A. 2004. *Ecotourism and community-based ecotourism in the Mekong region*, 43. Working paper no. 10. Sydney: Australian Mekong Resource Centre.

Li, W. 2006. Community decision making participation in development. *Annals of Tourism Research* 33(1): 132–143.

McGranahan, D. 1970. The interrelations between social and economic development. *Social Science Information* 9(6): 61–77.

Murphy, P.E. 1985. *Tourism: A community approach*. New York: Routledge.

Myrdal, G. 1974. What is development? *Journal of Economic Issues* 8(4): 729–736.

Nguyen.2010. = Viet Nam News. http://vietnamnews.vn/politics-laws/203923/president-addresses-development-summit-.html. Retrieved 12 Sept 2012.

Okazaki, E. 2008. A community-based tourism model: Its conception and use. *Journal of Sustainable Tourism* 16(5): 511–529.

Pfister, R., and Y. Wang. 2008. Residents' attitudes toward tourism and perceived personal benefits in a rural community. *Journal of Travel Research* 47(1): 84–93.

Prebisch, R. 1950. *The economic development of Latin America and Its principal problems (E/CN.12/89)*. New York: United Nations.

Preston, P.W. 1996. *Development theory: An introduction*. Cambridge, MA: Blackwell.

Pretty, J.N. 1995. Participatory learning for sustainable agriculture. *World Development* 23(8): 1247–1263.

Rostow, W. 1959. The stages of economic growth. *The Economic History Review* 12(1): 1–16.

Sa Pa District People Committee. 2013. *Year 2013 socio-economic, security and defence situation and orientation for 2014*. Lao Cai, Vietnam: Sa Pa District People Committee.

Salafsky, N., H. Cauley, et al. 2001. A systematic test of an enterprise strategy for community-based biodiversity conservation. *Conservation Biology* 15(6): 1585–1595.

Scheyvens, R. 1999. Ecotourism and the empowerment of local communities. *Tourism Management* 20(2): 245–249.

SNV (The Netherlands Development Organisation). 2007. *Community -based tourism: Lessons learnt from Viet Nam*. Ha Noi: Unpublished document.

Stake, R.E. 1995. *The art of case study research*. Thousand Oaks: Sage.

Stake, R.E. 2006. *Multiple case study analysis*. New York: The Guilford Press.

Strauss, A.L., and J.M. Corbin. 1990. *Basics of qualitative research: Grounded theory procedures and techniques*. Newbury Park: Sage.

Strauss, A.L., and J.M. Corbin. 1998. *Basics of qualitative research: Techniques and procedures for developing grounded theory*. Thousand Oaks: Sage.

Terlouw, K. 2001. Regions in geography and the regional geography of semiperipheral development. *Tijdschrift voor Economische en Sociale Geografie* 92(1): 76–87.

UN. 2000. *United Nations millennium declaration*. New York: United Nations, Department of Public Information.

UN. 2010. *The contribution of tourism to trade and development*. United Nations conference on trade and development, Geneva.

UNDP. 1990. *Human development report 1990*. New York.

UNDP. 2010. *Viet Nam at a glance*. http://www.undp.org.vn/. Retrieved 2 Sept 2010.

UNDP. 2012. Viet Nam profile of human development indicators. http://hdrstats.undp.org/en/countries/profiles/VNM.html. Retrieved 8 Sept 2012.

UNWTO. 2013. *UNWTO world tourism barometer*, 11. Madrid: UNWTO.

Viet Nam National Administration of Tourism. 2012. *Tourism statistics* [in Vietnamese]. http://www.vietnamtourism.gov.vn/english/index.php?cat=012040&itemid=5143

Wallerstein, I. 1974. *The modern world-system: Capitalist agriculture and the origins of the European world-economy in the sixteenth century*. New York: Academic.

Wallerstein, I. 1984. *The politics of the world-economy: The states, the movements, and the civilizations*. Cambridge, UK: Cambridge University Press.

Wanhill, S. 1997. Peripheral area tourism: A European perspective. *Progress in Tourism and Hospitality Research* 3: 47–70.

Weaver, D.B. 1998. *Ecotourism in the less developed world*. Wallingford: CAB International.

Weaver, D.B. 2006. *Sustainable tourism: Theory and practice*. Oxford: Elsevier Butterworth-Heinemann.

Weaver, D.B. 2010. Community-based tourism as strategic dead-end. *Tourism Recreation Research* 35(1): 206–208.

Weaver, D.B., and L.J. Lawton. 2008. Not just surviving, but thriving: Perceived strengths of successful US-based travel agencies. *International Journal of Tourism Research* 10(1): 41–53.

Weaver, D.B., and L.J. Lawton. 2013. Resident perceptions of a contentious tourism event. *Tourism Management* 37: 165–175.

Weaver, D.B., and L.J. Lawton. 2014. *Tourism management*. Fifth Edition. Milton, Qld: Wiley.

Wellhofer, E.S. 1989. Core and periphery: Territorial dimensions in politics. *Urban Studies* 26(3): 340–355.

World Bank. 2010. *Viet Nam country brief*. http://www.worldbank.org/en/country/vietnam/overview#outcomes. Retrieved 2 Sept 2010.

World Bank. 2012. *World development indicators*, 463. Washington, DC: The World Bank.

World Travel and Tourism Council. 2012. *Travel and tourism economic impact 2012 Viet Nam*. London, UK: The WTTC.

Xiao, H., and S.L.J. Smith. 2006. Case studies in tourism research: A state-of-the-art analysis. *Tourism Management* 27(5): 738–749.

Yin, R.K. 2009. *Case study research: Design and methods*. Thousand Oaks: Sage.

Zeppel, H. 2006. *Indigenous ecotourism: Sustainable development and management*. New York: CAB International.

Chapter 11 The Concept of Environmental Supply in National Parks

Barbara Jean McNicol

Abstract This study targets commercial tour operator's conceptualization of environmental supply in Canadian Rocky Mountain national parks. Environmental supply includes natural features, physical attributes, management initiatives and/or governance policies that influence environmental conservation. Environmental supply, therefore, will include the ecosystems of the park destination, visitor education and interpretive programs, zoning and access, environmental impact assessment (EIA) procedures, policies and guidelines, and any other characteristics and initiatives that impact the conservation and/or preservation of nature at a visitor-centered site. How commercial tour operators respond to and manage these for environmental protection in a conservation environment will determine tourism sustainability at the destination. This evaluation includes presentation of the results of 16 one-on-one field interviews from an illustrative sample of 85 commercial tour operators (managers) within Banff and Jasper national park boundaries (from June to November, 2011). Each interview included nine formal questions, ranging from a description of the types of activities included in tours to actual or perceived limitations placed on operators when locating within a national park. Three of the questions targeted specifically tour operator's understanding of environmental supply. Results are discussed and grouped for presentation.

Keywords National parks • Environment • Commercial tours • Sustainable tourism

11.1 Introduction

Over time, the role of tourism in natural areas has been upgraded and refocused to reflect ecological values and maintenance of the quality of the visitor experience. Sustainability, needed to ensure protection of the environment and quality tourism experiences, must engage ecological integrity while ensuring that visitors will be

B.J. McNicol (✉)
Mount Royal University, Calgary, Alberta, Canada
e-mail: bmicnicol@mtroyal.ca

S.F. McCool, K. Bosak (eds.), *Reframing Sustainable Tourism*,
Environmental Challenges and Solutions 2, DOI 10.1007/978-94-017-7209-9_11

informed about the importance of natural landscapes, cultural heritage, the role of conservation for society, and how a balance of these will be important for the future. These functions of sustainable tourism are a challenge in national parks where ecological values are often, in direct opposition to human use activities.

Environmental supply is a concept that integrates ecological considerations with tourism initiatives. The paradigm of environmental supply provides for communication and operation by integrating the goals of sustainable tourism with the values of ecological conservation. For example, ecosystem services include all of the natural and ecological systems within a natural environment while environmental supply, which operates within this framework, melds all aspects of visitor use to environmental management. Environmental supply is the supply of natural features, attributes, management initiatives and/or governance policies that influence the environmental conservation of tourism landscapes.

Sustainable tourism refers to a form of tourism in which the use of resources meets the needs of visitors while preserving the environment for the enjoyment of generations to come (Hall 1998). In this definition, sustainable tourism ties together concern for the carrying capacities of ecological systems with the challenges of human uses. The management of sustainable tourism includes environmental sustainability (environmental management) and sustainability of the tourism experience (visitor management) in national parks. Over time, an understanding of natural and ecological processes in national parks has evolved with a focus on physical scientific research. Concerns about appropriate environmental management initiatives have often evolved with disconnect from human use research and these factors have influenced environmental protection legislation and policies. The impacts of tourism activities in national parks, however, remain a threat to ecological integrity where visitor management objectives focus on commercial tourism ventures and development. Despite the best science, policy and legislative tools at their disposal, national parks such as Banff and Jasper, in the Canadian Rockies, continue to struggle with the consistent application of a balanced management approach between environmental protection and human uses (Swinnerton 2002 in Pavelka and Rollins 2009, p. 274). Since environmental supply is a concept that integrates the goals of environmental management with commercial and tourism considerations in natural areas, the purpose of this chapter is to introduce and evaluate the concept of environmental supply as it is perceived and interpreted by commercial tour operators in Banff and Jasper national parks. The chapter seeks to show where conservation of ecosystems and the sustainability of a quality visitor experience form the foundation of sustainable tourism within national parks.

11.2 The Concept of Environmental Supply

There is a need in federally protected areas, such as national parks, to investigate the connection between management actions and responsible and sustainable tourism. One way this can be accomplished is by linking commercial visitor demand with

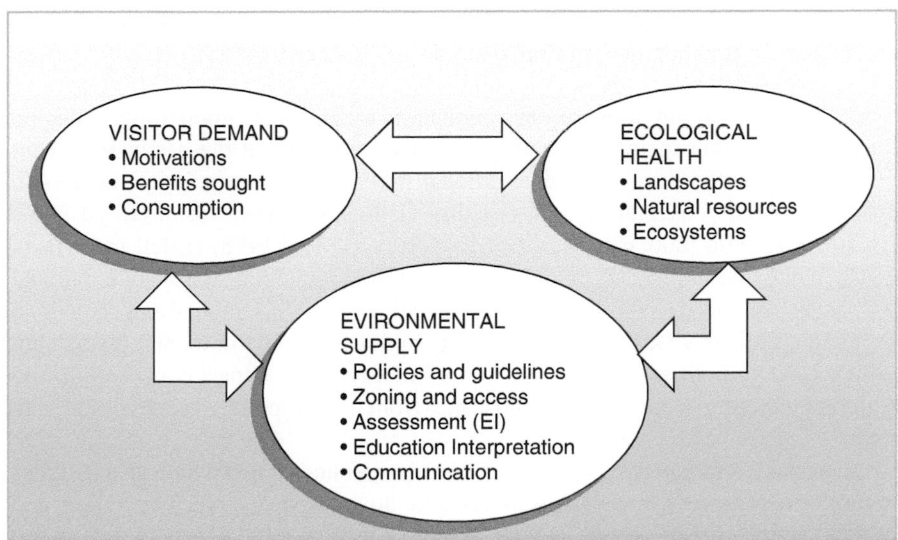

Fig. 11.1 Balancing visitor demand with environmental supply

environmental supply. The term *environmental supply* is a new one. It is a concept inherent in the planning and management of sustainable tourism yet it also is a concept that has defied specific definition and clarity until recently. The concept of environmental supply uses disconnect that occurs between visitor demand and what an ecosystem supplies to link management of human use to the conservation of natural areas. McNicol (2015, p. 210) has published the first official definition of environmental supply in the *Encyclopedia of Sustainable Tourism* compiled by CABI International.

Environmental supply operates at the interface of environmental and visitor management and requires integration of often opposing constructs such as visitor demand and ecological health (Fig. 11.1). Key to the concept of environmental supply is the inclusion of policies that place emphasis on the environment and conservation of ecosystems. Fundamentally, visitor demand is a form of consumer demand where people visit destinations to use and, often eventually consume natural products. In fact, Krippendorf (1976) famously presented an argument against the growing negative impacts of tourism on destinations by referring to tourists as 'landscape eaters'. He presented tourists as consumers of natural and cultural landscapes while eventually bringing about irreversible and negative changes to destinations targeted for tourism. Budowski (1976) also supported this representation by suggesting that tourism landscapes will reflect 'conflict or coexistence' dependent upon the behavior of visitors in contrastingly different environments than their own. While these are early abstractions of visitor and environmental interrelationships, these ideas have formed a foundation for questioning excessive visitor density and actions in natural tourism environments. Visitor impacts, with a focus on eventual degradation of natural landscapes, have been readily debated and emphasized in the tourism literature (Baretje 1977; Cohen 1978; Lime and Stankey 1979; McCool 1978).

A landscape, in any tourism environment, is a consumer product that relies on the sustainable management initiatives of selective supply at the destination (Hall 1998). A protected area, such as a national park, relies on sustainable actions to ensure resource conservation and a quality visitor experience result (Boyd 2002; Broson and Noble 2006; Honey 1999; Mowforth and Munt 1998; Tisdell and Wilson 2012; White and Noble 2012). The concept of environmental supply, in a national park, may be viewed as those initiatives that influence environmental and ecological protection such as visitor education, zoning and access, environmental assessment policies and guidelines (EA) and any other initiatives that impact conservation and support protection of natural park landscapes from inappropriate or excessive human uses.

Sustainable tourism is dependent upon environmental supply which also includes natural environments, natural resources and flow resources (such as water and energy) of the tourism system (Holden 2000). How tourism developers, commercial operators, planners and managers respond to and manage these for protection will determine tourism sustainability at the destination. Evaluating environmental supply is one facet of environmental planning and management that concentrates on environmental assessment and includes risk assessment as a way of understanding the consequences of visitor and environmental interactions on the natural landscape. Results include both an understanding of visitor safety as well as ecological sustainability. Research has evolved that integrates clear assessment of environmental management, quality performance and visitor management frameworks (for example, Heck et al. 2011; Moore et al. 2003) with emphasis in protected areas. No-where is a need for, and the concept of environmental supply, more applicable to sustainable tourism than in protected national park environments where visitor management and environmental management remain not only separate management divisions but separate research offices with, often, isolated and separately funded initiatives. For understanding of the sustainable tourism and environment nexus in protected areas, this has proven problematic. For example, in 2012 Banff National Park saw the end of the Social Science Research Division which was responsible for developing and implementing human use research in the Rocky Mountain Park system. This suggests that visitor and human use studies will be reduced for the immediate future.

Canada's national parks are designed and established to protect some of Canada's most spectacular yet fragile environments while at the same time promoting visitor accessibility and use. This paradox was ensured with proclamation of the National Parks Act in 1930. Canada's national parks have operated with a dual mandate; not only are national parks 'dedicated to the people of Canada for their benefit, education and enjoyment' but also parks are to be maintained and used so they will remain unimpaired for future generations (Canada Depository Services Program 2000). A dichotomy between park management emphasis placed on visitor management or ecological management has created conflict over time, generally between tourism and/or recreation advocates and environmentalists. A more modern perspective maintains it is time to place balance on integrating these co-existing positions. Understanding environmental supply, from stakeholder viewpoints, will contribute to this balance.

11.3 The Environmental and Visitor Management Integration

A park management emphasis directed on either ecological integrity or human use has tended to fluctuate over decades in national parks (Dearden and Berg 1993; Dearden and Rollins 2009). After a declared emphasis in 2001 on conservation of ecosystems, a change in 2009/2010 saw Banff National Park management focusing on increased visitation and recreation demand in the form of an announced 2 % increase annually over a 3 year period (Parks Canada 2010). Given that Banff and Jasper national parks are among two of the most 'enjoyed' national parks in Canada, there is a need to question whether this cumulative 6 % increase in visitation will test the abilities and availability of institutional and environmental supply. This important, and relatively recent shift in focus initiates a call for sound scientific information to guide decisions linked to re-investment in park facilities and programs that will promote positive visitor experience and enhance public awareness of the values and benefits of national parks. It is also clear that the bulk of new and diversified park recreation opportunities, such as via-ferratas or zip lines, will become activities offered by third party operations such as private commercial tour businesses. In fact, Mount Norquay, the local Banff ski hill, has had approved a 2013 proposal and has currently implemented a via-ferrata climbing route to increase summer usage and visitation to the local area. Using this as one example, balance between the demands for a quality visitor experience and the goals of maintaining ecological integrity is an ongoing paradox for the protected national park environment.

Environmental management and planning of visitor use in protected areas uses impact assessment as a way of understanding the negative consequences of human and environment interactions. Limits of Acceptable Change (LAC) and models such as Visitor Impact Management (VIM), that establish, or seek to predict, thresholds of carrying capacity, have a history of use in recreational and protected areas management (Holden 2000; Lime and Stankey 1979; Manning 2001; Stankey et al. 1985; Stankey and Schreyer 1985). Other models that focus on recreation, such as the Recreation Opportunity Spectrum (ROS), or tourism aesthetics, such as the Visual Landscape Inventory (VLI), provide information on existing recreation experience settings or information on the location of important viewing opportunities (viewpoints). Together these types of approaches provide a context for relative amounts of recreation use and user expectations as well as information on visitor expectations toward quality tourism experiences. There are a variety of practices that can be used to manage visitor use and associated environmental and resource impacts in natural environments, such as the Visitor Activity Management Process (VAMP) or Visitor Experience and Resource Protection (VERP) (Haider and Payne 2009; Manning 1999, 2007; Park et al. 2008). Depending on what visitor experiences are supplied, these forms of models provide a framework and process for defining limitations and creating guidelines, policies and regulations for visitor actions and access. Governments, both federal and provincial, have been at the forefront of implementing models based on maximum visitor use and examining diverse recreation opportunities. Forestry and parks departments, most concerned with outdoor recreation, have implemented many of these approaches over time (for example, Griest 1976; Stankey 1973; Stankey and Schreyer 1985; Vaske 1994).

In Canada's Rocky Mountain system of national parks, the approach used is a Human Use Management strategy (HUM) officially implemented in Banff and Jasper in 2004 (Kachi and Walker 1999; Parks Canada 2004, p. 42). Realizing also the predicaments created by relying solely on biology and other natural sciences, Parks Canada's latest corporate orientation documents have made an important shift to one that recognizes that ecological integrity cannot be achieved without 'people.' There is an increased emphasis upon human relations and interactions with nature and the involvement of citizens as partners and advocates for national park policy. A shift in focus of the most recent park management plans puts significantly more emphasis on visitor experience and personal connection to place than it has in the past; most evident in the Banff National Park management plan (Parks Canada 2004). Delivery of services and new activities will continue to rely on partnerships with third party commercial operators (Canada National Parks Act, Section 5 1998).

The premise of planning tools such as LAC, VIM or HUM, in a contemporary context, is that indicators can be developed that will predict and monitor for unacceptable visitor and environmental changes. These indicators allow environmental managers to work within an environment's carrying capacity and manage for dimensions of unacceptable changes. Social changes are linked to such indicators as 'overcrowding' or 'visitor satisfaction' and biophysical indicators such as 'species losses' or 'plant and wildlife mortality rates' are typical of high use recreational and tourism landscapes (Baral et al. 2012; Bossel 1999). In protected areas, where diversity of human recreational activities is a concern, the management focus has traditionally been on negative physical consequences measured by using biophysical indicators (Buckley 2000, 2004). The introduction of the concept of environmental supply, which incorporates the scientific term of 'environment', initiates a 'quality movement' in environmental management that proposes to switch the measurement of environmental indicators from those that focus on negative consequences of visitor use to those that place emphasis on environmental quality and success of the visitor experience. Being able to understand a positive visitor experience is one key characteristic of sustainable tourism. Being able to link a positive visitor experience to environmental consequences is important to the environmental management of national park conservation (Broson and Noble 2006; Tisdell and Wilson 2012).

Human attraction and use of natural environments, in the form of sustainable tourism, is a complex relationship that has been difficult to understand or conceptually model (West 2008). Some indicators are easier to quantify than others, such as biophysical or economic indicators based solely on numerical quantities. Environmental indicators of sustainable use that are visitor experience-based are more fluid and harder to measure (Bossel 1999; Buckley 2012). One recent emphasis in tourism and recreational research is on trying to determine and evaluate the harder to measure phenomena to ensure sustainability of both tourism and the environment (Moore et al. 2003). To gain understanding of sustainable landscapes in national parks the emphasis on creating and using indicators requires a balance between easier to measure biophysical indicators and the more problematic indicators that focus on, and measure, the quality of the visitor experience.

11.4 Banff and Jasper: Commercial Operators and Environmental Supply

The concept of environmental supply provides a framework for raising questions about all environmental aspects in a national park. The Four Mountain National Park system which overlaps the provinces of British Columbia and Alberta includes Kootenay, Yoho, Banff, and Jasper national parks. Banff and Jasper national parks are very similar for all but exact latitude (Fig. 11.2). As two of the most northern

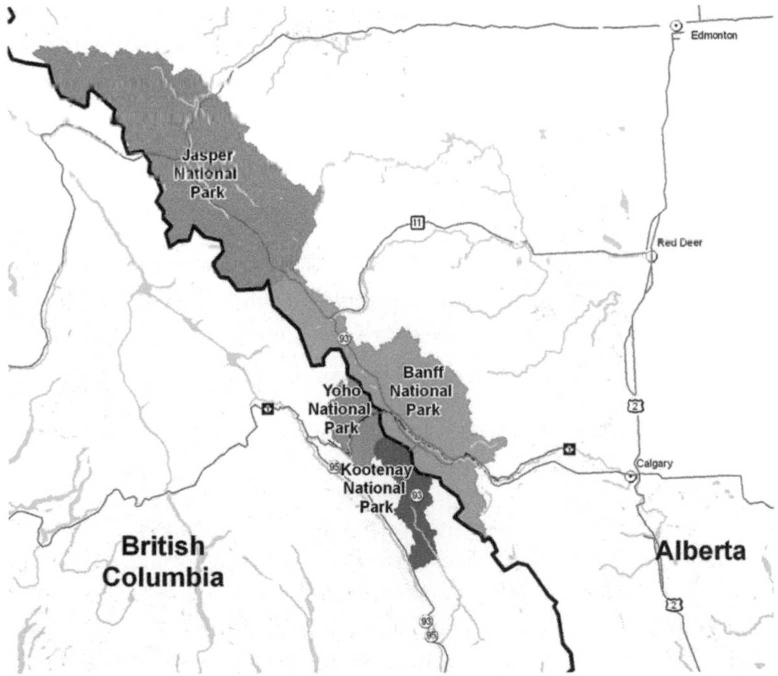

Fig. 11.2 Four Mountain Park System, Canadian Rocky Mountains

parks in the Four Mountain Park System of the Canadian Rocky Mountains these parks have similar management plans with congruent goals and objectives (Jasper National Park 2012; Parks Canada 2010). Banff and Jasper national parks, as the focus of a research area, encompass an integrated and shared natural history and are joined by cultural and heritage landscapes. For example, included within these parks is diversity of terrain, topography and ecosystems that are connected by human activities. Physical features include, the Central Rocky Mountain Ecosystem composed of Montane forests, vegetation and associated wildlife species in the Bow Valley of Banff National Park to the alpine and remnant glacial environments, such as the Columbia Ice Fields, found in Jasper National Park (McNicol 2009). Visitors to the northern Rocky Mountain region often visit Banff or Jasper, or both, and then continue directly west to Yoho or southwest to Kootenay national parks (Fig. 11.2).

Noting the similarities of these national parks, a study was undertaken to understand commercial tour operator's perceptions of the concept of environmental supply in Banff and Jasper national parks. The collective experience, knowledge, and insights of managers from destination management organizations, such as commercial tour operators, with expertise in destination management provide a valuable source of information within park boundaries (Crouch 2011). It is common to interview managers and other practitioners from public and private tourism sectors as this is the population that is the most knowledgeable about the tourism destination elements (Bossel 1999; Enright and Newton 2004). Key to this study is the clarification of the performance of commercial tour operators' (and client expectations) for activities and environmental experiences that national park management is able to supply. The field data gathered for this research project consisted of 16 one-on-one formal field interviews of tour operator managers. The field interviews, approved and supported by Parks Canada, were conducted from June to November, 2011, in both Banff and Jasper national parks.

Parks Canada provided a list of tour operators with business licenses that operated in Banff or Jasper or both national parks. These lists consisted of 60 businesses that had addresses located within the parks. In addition, the Banff and Lake Louise Tourism Bureaus were consulted and the Internet and phone books checked for possible enterprises missing from the original list. In the end, the final list consisted of 85 possible interviewees. It was discovered that over 2000 businesses, from all over the world, operated tourism itineraries in the parks but only 85 located their businesses directly within national park boundaries, usually in the Banff and Jasper town-sites.

The main goal of the field study was to understand tour manager's concept of environmental supply and gain an understanding of the role of environment in their commercial tourism activities. Questions were designed to understand who they were, when and where they were going within park boundaries, why they were using park landscapes, what activities they participated in, their understanding of environmental policies, and how they were defining environmental supply within these national park environments as this pertained to the quality of business operations and commercial visitor experiences. Nine key questions were asked of each interviewee within a semi-structured and open-ended format. The questions were:

1. What recreation activities are you offering?
2. What countries do your clientele represent?
3. What locations in the national park do you visit and why?
4. What types of environmental supply characteristics are important for the quality of running your business?
5. What advantages are there of the class screening process to your commercial operation?
6. What disadvantages are there of the class screening process to your commercial operation?
7. What aspects of the ecosystem are most important for a successful activity by your tour company?
8. What aspects of park management are the most important for a successful activity by your tour company?
9. Is there anything else you would like to discuss about conducting (a) quality visitor experiences and (b) successful commercial tourism operations in Banff (or Jasper) National Park?

The interviews took on average 20 min to complete and ranged from 10 to 45 min. Interviewees had difficulty with some of the terminology, especially with questions 4 and 7. The definitions of the terms 'environmental supply' and 'aspects of the ecosystem' were provided to the interviewees on a piece of paper so that they could read the definitions themselves. These field interviews were confidential for the interviewee. Responses were presented as descriptive and anonymous statements.

The interviews were transcribed and responses were classified into categories based on key terminology. For the purposes of this paper, the key questions, specific to commercial operators understanding and perception of environmental supply, were questions 4, 7 and 8. These questions determined environmental supply characteristics, management aspects and ecosystem characteristics which are catalogued and grouped into categories in Table 11.1. Grouped responses are prioritized in Table 11.1 to show the number of times each terminology as a 'characteristic' or 'aspect' was mentioned in a response. Responses that were mentioned only once, solely by a single commercial operator, were included in Table 11.1 to show the complete range of responses. Questions 5 and 6 of the field interviews were designed and included to inform the process of class screening, which is an important policy and applied aspect of environmental assessment, and hence environmental supply in Banff and Jasper national parks. These two specific questions were included since Parks Canada requires indicators of sustainability to use for evaluating the assessment of class screening effectiveness by commercial tourism operators (Heck et al. 2011; Jamal and Dredge 2011; Tribe 2008). Categories of responses from these two questions provide a foundation upon which to understand further and future monitoring by tourism businesses of the performance of class screening processes in national parks.

Table 11.1 Environmental supply defined by commercial tour operators in Banff and Jasper National Parks (times mentioned)

Environmental supply characteristics	Management aspects	Ecosystem characteristics
Education (11) Zoning and access (10) Ecological protection (5) Group or equipment limits (3) Park interpretation (2)	Warden services (3) Product development (3) Balance of environment and business (3) Trial maintenance (3) Wildlife management (3) Park management plan (2) (as it addresses the visitor experience) Infrastructure (2) Growth for businesses (2) Communication of park goals(2)	Wildlife (15) Scenery (10) Water cycle (6) Flora (4) Weather (4) Wilderness (3)
(1) Safety Fishing regulations Quality experiences Sustainable business practices Long-term sustainability Limits on itineraries Impact assessment	**(1)** Access management **(2)** Gate fees Promotion and marketing Enforcement of certification Ecosystem management Product delivery Education Signage Road maintenance Stewardship Budget Multi-use trail designation High usage development Fish stock management Ski operations Diversity of business activities	**(1)** Pollution in water Access to ecosystems Glaciers Insects Water fowl Wildlife habitat

11.5 Commercial Tour Operators Define Environmental Supply

The data collected from the one-on-one interviews produced interesting results about commercial tour operations in Banff and Jasper national parks. Hiking, interpretive tour guiding and wildlife viewing were the leading activities that the tour operators provided. There were, however, a diverse range of activities that also included fishing activities, canoe rentals, horseback riding, snowboarding, snow shoeing, rafting, gondola rides, interpretive centre visits, cross-country skiing, backcountry accommodations and small boat rentals. As the interviews progressed it became apparent that the type of recreational activities offered by companies tended to influence responses about environmental supply; fishing tour operators' responses would concentrate on aquatic management while concerns about horse packing tours would concentrate on land-based management issues. For example, one tour company offering fishing experiences to tourists stated that their understanding of environmental supply, in their national park, was linked directly to "limits on tour boat numbers, regulations on the types of motors that can be used, limits to the number of rentals allowed, and the need for environmentally friendly boats". On the other hand, a company offering overnight back country horse packing trips discussed environmental supply as "restrictions on travel through the back country, wildlife protection, a need for education about wilderness activities, and the addition of quality to their visitor's experience". Environmental supply, for individual tour operators, was clearly linked to the types of recreational activities and offerings by the company which included recognition of limitations and regulations of park access and actions.

Canadians, Americans and visitors from the United Kingdom (predominantly those from England) were the main nationalities that use the tour operations within Banff and Jasper national parks. Commercial clientele, however, were present from all countries of the developed world with many from Australia, Germany, Japan, Korea and China. India, Mexico, Brazil, Denmark, Russia and visitors from Holland were mentioned solely as the clientele of singular companies. It was determined that the tour businesses geographically located their tourism experiences throughout the parks with a higher concentration of tour operators taking clientele to (1) the Columbia Ice Field in Jasper National Park, (2) Lake Minnewanka in Banff National Park, and (3) Maligne Lake in Jasper National Park. Access to park landscapes, based on the type of zoning use, was an ongoing theme when discussing locations of operations and park destinations that tour operators included in their itineraries. For example, horse-use trail designations predicted where backcountry horse riders would go in the park and, in general, the tours went to "areas of interest to the customer which have allocated permits" or "where hikes are allowed with designated group size". A number of respondents suggested that some trails should become multi-use and that limited accessibility designated by zoning was an access concern. It was not surprising therefore, that access and zoning were characteristics of environmental supply mentioned by ten of the operators interviewed (Table 11.1).

Characteristics of environmental supply that were deemed the most important for the running of the tour businesses were education, zoning and access, ecological protection, limits on group size or equipment uses, and park interpretation services (Table 11.1). The identified characteristics of environmental supply are important to business success and quality visitor experiences while operating within a Canadian national park environment (Boyd 2002; Hvengaard et al. 2009; Patterson 2007). Hvengaard et al. (2009) suggest that education and park interpretation services provide information and visitor awareness which are important aspects of environmental conservation. Patterson (2007) noted also, that for tour operations, longer-term and, often costly, marketing budgets are important for nature-based tour guiding operations since there is a need to incorporate educational images and conservation awareness into the processes. The necessity of class screening standards within mountain parks may also have created a perception of an increased need for educational activities and use of park interpretive services (McNicol and Draper 2010). In other words, operating within a national park will produce different demands on tour operators than on those conducting nature-based businesses outside of national park boundaries.

In both the environmental and tourism literature the ability to regulate numbers and to designate use in sensitive and critical environments is deemed important (for example: Buckley 2004; Holden 2000; Tisdell and Wilson 2012). From an environmental management perspective, this means that the geographical and spatial designation of park use ensures that itineraries do not include sensitive conservation areas, such as wildlife corridors or critical animal reproductive habitats, in daily schedules. From a commercial operator's perspective, this often includes an ecological aspect since 'ecological protection' was mentioned as an important characteristic of environmental supply in at least 5 of the 16 interviews (Table 11.1). It is clear that the businesses interviewed had understanding of their environmental commitments while operating within a protected area. One commercial tour operator suggested that there should exist "more opportunities to create awareness around preservation of the park" and another stated that "business growth should provide opportunity to create an experience without compromising ecological integrity within the park". Discussion during the interviews indicated that conservation and ecological integrity were very important to most commercial tour operators conducting tours in Banff and Jasper national parks.

The most important aspects of park management for the businesses were the park warden service (which plans and monitors natural resource management), developing new products within the park, and management that can create a balance between environment and businesses (Table 11.1). More specific concerns, were trail and wildlife management, which were also reflected in the important environmental supply categories of 'access and zoning' and 'ecological protection' (Table 11.1). Wildlife viewing is very important to a quality tourism experience in mountain parks and managers are clear that access to wildlife and management of wildlife is important for their businesses. Development of new recreational products also was perceived of importance. Referencing the current Banff National Park management plan (2010), park managers announced a push for increased visitation and diversification of recreational opportunities in Canadian mountain parks such as Banff and Jasper (Parks Canada 2010). New projects, such as the glass Brewster Discovery Walk and the Mount Norquay via ferrata, have provided new entry-level adventures for park visitors and new product development in the form of recreational opportunities for clients. Four interviewees directly stated "the need for new activities within the parks" and four operators mentioned that "corresponding increases in infrastructure is needed to help (repeat) visitation grow" (Table 11.1).

Greater emphasis on developing new businesses in the parks and helping existing ones grow (within the context of increasing visitor numbers and demands) was important to a number of operators. Of note, perceived differences between older and newer businesses and differences of educational levels of tour company managers were voiced by two companies as points of concern. For example, the manager of a fishing tour company, with a Masters degree, expressed concern about a lack of tour company input into Parks Canada's fisheries planning and management in the mountain parks. The tour operator suggested that "they [Parks

Canada] make resolutions before asking for business operator's opinions within the process". Communication and dialogue between Parks Canada managers and commercial tourism businesses was a key park management issue that arose during discussions linked to question 9.

The age or 'longevity' of a tour business operating within either Banff or Jasper was a concern that was emphasized repeatedly. For example, Brewster's Inc. (Guiding Company) was established in 1892, not long after park designation in 1885, and continues to operate a diversity of tourism opportunities within both Banff and Jasper national parks. Other, newer, usually smaller businesses must now undergo class screening and establish themselves within a hierarchy of well-established older tour businesses. Older companies expressed a need for "better screening processes for newer companies" while newly establishing businesses suggested that "smaller businesses are hurting" and finding operating difficult "to compete with tour businesses from outside the park". The previous comment refers to the many newer commercial tour operators that have been encouraged by Parks Canada policies to establish outside national park boundaries, developing operations in neighboring towns and cities, and transporting tourists in and out of the parks on a regular basis. These external companies are not required to meet the same demands as companies locating within a Canadian national park.

The most significant aspects of the ecosystem for the running of the businesses were: wildlife, scenery, the water cycle, weather, and wilderness (Table 11.1). These are most often the main (general) responses given when stakeholders are questioned about tourism and ecology in national parks. These are important, but typical, and limiting if they are taken to be the only environmental indicators of sustainable tourism within national parks. As addressed throughout this article, the concept of environmental supply expands upon the scientific definition of environment to integrate all aspects of visitor use and the environment. Environmental supply is much more comprehensive than solely (and generic) ecosystem characteristics and includes not only natural features and physical attributes but management initiatives and/or governance policies that influence environmental conservation. Within the context of Banff and Jasper national parks it is better possible to understand the interactions of visitor use and commercial tourism operations as reflections of aspects and elements of environmental supply.

11.6 Re-framing Sustainable Tourism

Sustainable tourism should include the concept of environmental supply. The integration, between sustainable tourism and environmental management occurs at the tourism and environment interface where conservation is a key theme of emphasis (Honey 1999; Tisdell and Wilson 2012). At the same time, how commercial tourism operators perceive their roles and abilities within national park environments is key to achieving balance between competing environmental and recreational objectives. Nature-based tourism businesses within national parks, which are essentially economically-centered, are aware of a need to sustain the natural resources and ecological values that attract clients to the park (Jamal and Dredge 2011; Tisdell and Wilson 2012). To communicate, an understanding of park 'environment' requires a comprehensive understanding of ecosystems, policies, education, and management initiatives.

Comprehensive discussion with commercial tour operators, during field interviews in Banff and Jasper national parks in the Canadian Rocky Mountains, defined perceptions of environmental supply. At the same time, tour manager priorities in Banff and Jasper may prove different from other stakeholders in other national parks and in other geographical locations. Commercial guiding activities require standards that ensure that visitor and environmental conservation goals are being met (Buckley 2012) and these may prove different from policies and assessment standards required elsewhere. They may also prove different from the operation of sustainable tourism initiatives in forestry land use designations or other recreational areas, and especially from tourism resort destinations that are implementing 'green' or expected environmental standards in high-use landscapes. Applying the concept of environmental supply, however, ensures that consideration is given to integration of environmental management and human use management in all natural environments where policies for environmental protection are keyed to conservation. Defining differences of visitor demand and commercial needs may prove more important for sustainable tourism outcomes than standardizing similarities and regulations over resource sector systems, or even geographical regions and tourism destinations.

Sustainability of protected mountain environments, with high visitation and a diversity of tourism activities, includes recognition by national park stakeholders of the need to balance commercially hosted tourist behaviors with resource protection and conservation initiatives (Jamal and Dredge 2011; Manning 2001). Skipping an important step of how environmental supply is interpreted by suppliers as they seek to accommodate tourist demand suggests that researchers (and tourism managers) often miss a clear understanding of what needs to be measured for integration of visitor management with environmental management (McCool et al. 2001; Ryan 2003; Tribe 2008). Observations from this study can be used to designate categories of environmental supply and create indicators of sustainability, and build upon existing data bases, for further understanding and longer term input into natural area conservation in national parks. While traditional approaches of LAC and VAMP have provided past frameworks for human use management (Eagles and McCool 2002), understanding of environmental supply allows for specific actions and use of management tools, such as zoning, group size limits, and limits to access, to be identified, implemented and monitored. It provides a means for designating what types of management tools might be useful in an environmental risk situation. For example, in Banff National Park, adaptive management processes around human use and grizzly bear encounters at Moraine Lake in Lake Louise have included limitations to group size and seasonal trail closures with enhanced communications between park administration and both the park public and commercial tour groups about changing access and policies. Results from this study, as presented in Table 11.1, supports this need for increased communication of park goals to tourism operators, especially as it addresses quality of the visitor experience of those seeking a group tour product.

In order to function effectively within a national park, commercial operators need to know and understand park management directives, policies and operational procedures. If there is disconnect, then both parties need to understand where disconnect occurs in order to achieve a balance of goals and objectives that will result in sustainable tourism. Integration of sustainable nature-based tourism and environmental management, using the concept of environmental supply, is important for the future protection of Canada's Rocky Mountain national parks and the sustainability of critical natural environments in general.

References

Baral, N., M.J. Stern, and A.L. Hammett. 2012. Developing a scale for evaluating ecotourism by visitors: A study in the Annapurna Conservation Area, Nepal. *Journal of Sustainable Tourism* 20(7): 975–989.

Baretje, R. 1977. Tourist carrying capacity. *Essai Bibliographiques, Essais 1*. Aix-en-Provence: Cente des Hautes Etudes Touristiques.

Bossel, H. 1999. *Indicators for sustainable development: Theory, methods, applications*. Winnipeg: International Institute for Sustainable Development.

Boyd, S.W. 2002. Tourism, national parks and sustainability. In *Tourism and national parks, issues and implication*, ed. R.W. Butler and S.W. Boyd. New York: Wiley.

Broson, J., and B. Noble. 2006. Measuring the effectiveness of Parks Canada's environmental management system. *The Canadian Geographer* 50(1): 101–113.

Buckley, R. 2000. Tourism in the most fragile environments. *Tourism Recreation Research* 25: 31–40.

Buckley, R. 2004. Impacts positive and negative: Links between ecotourism and environment. In *Environmental impacts of ecotourism*, ed. R. Buckley. New York: CABI.

Buckley, R. 2012. Sustainability reporting and certification in tourism. *Tourism Recreation Research* 37(1): 85–90.

Budowski, G. 1976. Tourism and environmental conservation: Conflict, coexistence, or symbiosis? *Environmental Conservation* 3(1): 27–32.

Canada Depository Services Program (DSP). 2000. Parliamentary Research Branch, Bill C-27: *Canada National Parks Act*, by John Craig, LS-365E, Ottawa. http://dsp-psd.pwgsc.gc.ca/Collection-R/LoPBdP/LS/362/c27-e.htm

Canada National Parks Act. 1998. National parks of Canada businesses regulations. *National Parks of Canada Businesses Regulations*, Section 5, SOR/98-455, Ottawa.

Cohen, E. 1978. The impact of tourism on the physical environment. *Annals of Tourism Research* 5: 215–237.

Crouch, G.I. 2011. Destination competitiveness: An analysis of determinant attributes. *Journal of Travel Research* 50(1): 27–45.

Dearden, P., and Lawrence Berg. 1993. Canada's national parks: A model of administrative penetration. *Canadian Geographer* 37: 194–211.

Dearden, P., and R. Rollins. 2009. Parks and protected areas in Canada. In *Parks and protected areas in Canada: Planning and management*, 3rd ed, ed. P. Dearden and R. Rollins, 3–23. Toronto: Oxford University Press.

Eagles, P., and S.F. McCool. 2002. *Tourism in national parks and protected areas, planning and management*. New York: CABI.

Enright, M.J., and J. Newton. 2004. Tourism destination competitiveness: A quantitative approach. *Tourism Management* 25(6): 777–788.

Griest, D.A. 1976. The carrying capacity of public wildland and recreation areas: Evaluation of alternative measures. *Journal of Leisure Research* 8: 123–128.

Haider, W., and R.J. Payne. 2009. Visitor planning and management. In *Parks and protected areas in Canada: Planning and management*, 3rd ed, ed. P. Dearden and R. Rollins, 169–201. Toronto: Oxford University Press.

Hall, C.M. 1998. Historical antecedents of sustainable development and ecotourism: New labels on old bottles? In *Sustainable tourism, a geographical perspective*, ed. C.M. Hall and A.A. Lew, 13–24. New York: Addison Wesley Longman Limited.

Heck, N., P. Dearden, A. McDonald, and S. Carver. 2011. Stakeholder opinions on the assessment of MPA effectiveness and their interests to participate at Pacific Rim National Park Reserve Canada. *Environmental Management* 10: 1–24.

Holden, A. 2000. *Environment and tourism*. London: Routledge.

Honey, M. 1999. *Ecotourism and sustainable development, who owns paradise*. Washington, DC: Island Press.

Hvengaard, G.T., J. Shulis, and J.R. Butler. 2009. The role of interpretation. In *Parks and protected areas in Canada: Planning and management*, 3rd ed, ed. P. Dearden and R. Rollins, 202–234. Toronto: Oxford University Press.

Jamal, T., and D. Dredge. 2011. Editorial, certification and indicator. *Tourism Recreation Research* 36(3): 1.

Jasper National Park. 2012. *Jasper National Park annual report*. Jasper: Parks Canada.

Kachi, N., and K. Walker. 1999. *Status of human use management initiatives in Parks Canada*. Hull: Ecosystems Management Branch, Parks Canada.

Krippendorf, J. 1976. *Die landschaftsfresser. tourismus und erholungslandschaft – Verderben oder segen?* Bern: Hallwag.

Lime, D.W., and D.H. Stankey. 1979. Carrying capacity: Maintaining outdoor recreation quality in land and leisure. In *Concepts and methods in outdoor recreation*, ed. C.S. Van Doren et al., 105–118. Chicago: Maaroufa Press.

Manning, R.E. 1999. *Studies in outdoor recreation: Search and research for satisfaction*, 2nd ed. Corvallis: Oregon State University Press.

Manning, R.E. 2001. Visitor experience and resource protection: A framework for managing the carrying capacity for national parks. *Journal of Park and Recreation Administration* 19(1): 93–108.

Manning, R.E. 2007. *Parks and carrying capacity, commons without tragedy*. Washington: Island Press.

McCool, S.F. 1978. Recreation use limits: Issues for the tourism industry. *Journal of Travel Research* 17(2): 2–7.

McCool, S.F., R.N. Moisey, and N.P. Nickerson. 2001. What should tourism sustain? The disconnect with Industry perceptions of useful indicators. *Journal of Travel Research* 40(2): 124–131.

McNicol, B. 2009. National parks as outdoor classrooms: Environmental and tourism curriculum in Banff and Jasper national parks. *Research in Geographic Education* 10(1): 48–59.

McNicol, B. 2015. Environmental supply. In *The encyclopaedia of sustainable tourism*, ed. C. Cater, B. Garrod, and T. Low, 210. Wallingford: CABI.

McNicol, B., and D. Draper. 2010, *Staying commercial in National Parks*. Paper presented at the tourism and entrepreneurship conference, NEXT Institute, School of Business and Economics, Wilfrid Laurier University, Kichener-Waterloo.

Moore, S., A.J. Smith, and D.N. Newsome. 2003. Environmental performance reporting for natural area tourism: Contributions by visitor impact management frameworks and their indicators. *Journal of Sustainable Tourism* 11(4): 348–375.

Mowforth, M., and I. Munt. 1998. *Tourism and sustainability, new tourism in the third world*. New York/London: Routledge.

Park, L.O., R.E. Manning, J.L. Marion, S.R. Lawson, and C. Jacobi. 2008. Managing visitor impacts in parks: A multi-method study of the effectiveness of alternative management practices. *Journal of Park and Recreation Administration* 26(1): 97–121.

Parks Canada. 2004. *Banff National Park management plan, introduction to amended management plan*. Banff National Park: Parks Canada Agency.

Pavelka, J., and R. Rollins. 2009. Case study: Banff and Bow valley. In *Parks and protected areas in Canada, planning and management*, 3rd ed, ed. Philip Dearden and Rick Rollins, 272–292. Don Mills: Oxford University Press.

Parks Canada. 2010. *Banff National Park management plan*. Banff National Park,: Parks Canada Agency.

Patterson, C. 2007. *The business of ecotourism: The complete guide for nature and culture-based tourism operators*. Victoria: Trafford.

Ryan, C. 2003. *Recreational tourism: Demands and impacts*, Aspects of tourism. North York: Channel View.

Stankey, G.H. 1973. *Visitor perception of wilderness recreation carrying capacity*. Research paper INT-142, Intermountain Forest and Range Experimental Station, USDA Forest Service, Ogden.

Stankey, G.H., and R. Schreyer. 1985. Attitudes toward wilderness and factors affecting visitor behavior: A state-of-knowledge review. *National wilderness research conference: Issues, state of knowledge and future directions*, 246–293. USDA Forest Service, Inner-mountain Research Station, General Technical Report, INT-220. Ogden, UT.

Stankey, G.H., D. Cole, R. Lucas, M. Peterson, S. Frissel, and R. Washburne. 1985. *The limits of acceptable change (LAC) system for wilderness planning*. Department of Agriculture, USDA Forest Service, General Technical Report, INT-176. Washington, DC.

Tisdell, C., and C. Wilson. 2012. *Nature-based tourism and conservation, new economic insights and case studies*. Cheltenham/Northampton: Edward Elgar.

Tribe, J. 2008. Sustainability indicators for small tourism enterprises – An exploratory perspective. *Journal of Sustainable Tourism* 16(5): 575–595.

Vaske, J.J. 1994. *Social carrying capacity at the Columbia Ice field: Applying the visitor impact management framework*. Ottawa: Department of Canadian Heritage, Parks Canada.

West, P. 2008. Tourism as science and science as tourism. *Current Anthropology* 49(4): 597–626.

White, L., and B.F. Noble. 2012. Strategic environmental assessment for sustainability: A review of a decade of academic research. *Environmental Impact Assessment Review*. doi:10.1016/j.eiar.2012.10.003

Chapter 12
Sustainable Tourism in Brazil: Faxinal and Superagui Case Studies

Jasmine Cardozo Moreira, Robert C. Burns, and Valéria de Meira Albach

Abstract The tourism experience in Brazil is distinctive in many ways. In this chapter we discuss two case studies in two very different settings, but both located in the southeastern Brazil state of Paraná. Both emphasize the need for economic benefit to communities relying on tourism. The first focuses on the Faxinal and its contribution as a micro level sustainable tourism location. The second study involves the island residents near Superagui National park. This ecotourism setting includes extremely rare species and some limited economic impacts. Both case studies emphasize the importance of managing ecosystems and engaging local communities.

Keywords Sustainable tourism • Traditional communities • Local products and services • Brazil

12.1 Introduction

The nation of Brazil has recently undergone major transformation, moving from a newly formed democracy in the mid-1980s to an emerging world economic power in 2015. With over 215 million inhabitants and a land area larger than the continental United States, Brazil promises to be a key player in South American natural resource recreation in the coming years and decades. With a rich, diverse supply of natural resource settings, combined with two mega-events (2014 FIFA World Cup and 2016 Summer Olympics) Brazil is poised to become a destination choice of many tourists.

J.C. Moreira • V.de M. Albach
Departamento de Turismo, Ponta Grossa State University, Pca Santos Andrade, 01, Ponta Grossa, PR 84100-000, Brazil
e-mail: jasmine@uepg.br

R.C. Burns (✉)
School of Forestry and Natural Resources, West Virginia University, 6125 Percival Hall, Morgantown, WV 26501, USA
e-mail: robert.burns@mail.wvu.edu

S.F. McCool, K. Bosak (eds.), *Reframing Sustainable Tourism*,
Environmental Challenges and Solutions 2, DOI 10.1007/978-94-017-7209-9_12

The World Tourism Organization expects that tourism to Brazil will double over the next 20 years, with these two sporting mega-events serving as major marketing vehicles.

The tourism experience in Brazil is unique in many ways. Like many great tourism destinations, the very name "Brazil" often evokes a pre-conceived notion. Brazil is unique in that it provides such a vast variety of tourism opportunities, across a broad spectrum of tourism settings. Conversely, a tourist may select the most remote possible settings in which to recreate, from the rivers of the Amazon to remote, wilderness national park settings, to vast stretches of intensely beautiful beaches, devoid of tourists and human impacts. While many people around the world focus on the amazing mass tourism opportunities and settings available in Brazil (e.g., Carnival, developed beach tourism, etc.), this paper describes two unique nature-based tourism settings – and their role as examples of sustainable ecotourism at a micro level.

Brazil is unique in its political and economic realities as well. Although a rising middle-class society has emerged over the past decade, many social problems, such as the continued existence of extremely poor and disenfranchised people, will vex Brazil for decades to come. High intensity agriculture and resource extraction are critically important sectors of the Brazilian economy, with many state-owned or state-supported entities engaged in extractive businesses. Brazil has been plagued by land ownership problems, often related to timber extraction, agricultural interests and their interactions with local populations, including indigenous peoples.

Of critical importance is *how* Brazil's increase in tourism will occur. Brazil has linked itself closely with mass tourism, and to a lesser degree, nature-based or sustainable tourism. The role of sustainable tourism in developing countries, including Brazil, cannot be understated. Tourism unchecked may bring with it as many problems as it brings jobs and benefits to local communities, which would certainly not be considered sustainable tourism.

UNESCO defines sustainable ecotourism as tourism that respects both the local people and the traveler, cultural heritage and the environment. Accordingly, sustainable tourism falls under the four UNESCO dimensions of sustainable development; conservation, appropriate development, democracy, and peace, equality and human rights, across the four spheres of natural, political, economic and social. Other definitions of sustainable tourism are used frequently, as are definitions of responsible tourism, ecotourism, geo-tourism, volun-tourism, and so on. As the editors expressed in an earlier section of this book, agreement about the definition of sustainable tourism development is unlikely, and we need to ask ourselves which definition is most appropriate for a developing nation such as Brazil.

What is clear is that Brazil needs sustainable tourism, regardless of the nuances of the various definitions, and that Brazil's natural resource settings have the capacity to absorb more tourism. The Chico Mendes Institute (ICMBio) is the federal government entity responsible for parks and protected areas within Brazil. With support from the US Agency for International Development (USAID) and US Forest Service International Programs (USFS-IP), ICMBio outlined a sustainable tourism plan in support of the 2014 FIFA World Cup. Ten "Parques da Copa" (World Cup Parks) were identified near the 12 cities across Brazil in which the FIFA soccer games were held. This mega-event required an immense financial investment into protected area infrastructure, including a substantial amount into the Parques da Copa (Palhares 2012). The 2016 Summer Olympics, as well, will be held in Rio de Janeiro, furthering Brazil's image as a tourism destination.

In a country with massive poverty and other social problems, it is fair to ask if the investment into two of the world's largest and most prestigious sporting activities will result in a sustainable tourism base for Brazil. Only time will tell if the investment will result in a long term increase in sustainable tourism within Brazil.

Even with great strides recently, sustainable tourism development in Brazil has seen limited success. Although federal tourism policies have been implemented over the past two decades, with positive outcomes, results have been quite varied. One of the earliest policies promoting tourism was the 1994 Program for Tourism Development, which decentralized tourism policies and encouraged local and regional tourism efforts. In 1996, Brazil implemented its National Tourism Policy (PNT), which also focused on local and regional tourism planning efforts. This was followed by the 2004 National Program for the Regionalization of Tourism, a continuation and refinement of earlier efforts. The crux of this effort was on tourism development at a regional level, with an emphasis on local products, nature, and culture (Araujo and Dredge 2012).

12.2 Defining Sustainable Tourism

The concept (and even the definition) of sustainable ecotourism has been hotly debated since its early inception, and this debate continues today. Wall (1997) posited a simple, three word question that we are still attempting to answer today "Is Ecotourism Sustainable." To understand why this question still remains largely unanswered, a review of the literature related to ecotourism and sustainable tourism is necessary.

Blangy and Wood (1993) suggested sustainable ecotourism includes responsible travel to natural areas, the conservation of the environment, and the sustainability of local peoples. In an effort to bridge the perceived gap between theory and practice, Ross and Wall (1999) operationalized ecotourism as a strategy to support conservation and supporting local development. This effort sought to develop a framework with objectives and indicators against which ecotourism could be measured. These included protection of environment, education, financial gain, local participation, and quality tourism. Boyd and Butler (1996) made an early attempt to examine ecotourism as a variation of an existing social carrying capacity procedure the Recreation Opportunity Spectrum (ROS) called the Ecotourism Opportunity Spectrum (ECOS). The paper was a reaction to a steep increase in ecotourism in the mid-1980 to mid-1990s; a realization that the management of ecotourism was a necessary step in the evolution of the concept of ecotourism.

Table 12.1 Typology of ecotourism studies

Study	Description of content
Alaeddinoglu and Can (2011)	Nature-based tourism resources in Lake Van basin, Turkey
Alam et al. (2009)	Sustainable forest-based tourism in Bangladesh
Buultjens et al. (2003)	Sustainable forest-based tourism in South Wales, Australia
Gouvea (2004)	Latin American ecotourism challenges and opportunities
Lai and Nepal (2006)	Local perspectives of ecotourism development in Tawushan Nature Reserve, Taiwan
Navalporto et al. (2012)	An examination of tourism development in Spain national parks
Reimer and Walter (2013)	Community-based tourism in Cardamom Mountains of southwestern Cambodia
Torres-Sovero et al. (2012)	Tourist satisfaction in Peruvian Amazon ecotourism lodges
Tsaur et al. (2006)	Taiwanese indigenous community tourism resources
Wright (1995)	Brazil sustainable ecotourism: balancing economic, environmental and social goals within an ethical framework
Wunder (2000)	Ecotourism and economic incentives to local indigenous people in Ecuadorian Amazon region

ECOS focused primarily on the experience sought and provided by ecotourism settings and providers. While this procedure allowed academicians and practitioners to reference other social carrying capacity procedures (e.g., Recreation Opportunity Spectrum (ROS), Limits of Acceptable Change (LAC), and others), one glaring omission was a complete lack of focus on local development or income for local peoples.

The importance of including an economic variable in discussing ecotourism has been heavily emphasized in plethora of publications over the past two decades. Hill and Gale (2009) clearly made the point that economic issues must be considered (e.g., the triple-bottom-line-sustainability (TBLS) concept (Hill and Gale 2009), ecotourism as an economic force in developing communities (Campbell 1999; Gouvea 2004; Silva and McDill 2004; Navalporto et al. 2012; Wunder 2000).

That this economic impact has the potential to be felt in a given local setting has been expressed extensively in the study of ecotourism and sustainable travel and tourism. Case studies have been used extensively, focusing on local and indigenous peoples in developing settings where ecotourism has taken place. Table 12.1 (above) demonstrates some of the varied settings and foci of ecotourism case studies germane to the case studies described in this chapter.

Common themes are seen through the majority of these manuscripts, regardless of continent, nation or community. These include the role of the local people, economic impact/financial impacts, care for the environmental resources, and often engagement of local agents, such as governmental or non-governmental organizations.

12.3 Brazil Tourism

It is generally accepted that all forms of tourism should strive to be sustainable in today's society, regardless of the continent or nation. Sisto (2003) suggested that sustainability, a premise developed over the past two decades, is now a universally accepted practice, and the term is often used by contemporary politicians and scholars when discussing tourism. In other words, the term sustainable tourism has become ubiquitous when discussing nature-based tourism. Another concept that has been relatively universally accepted is that there is an economic cost for tourism that is considered sustainable. The cost, borne by tourists, is a necessary component of sustainable tourism that allows special places to remain special. Such economic costs can also provide for better tourism experiences, which is also a positive attribute of sustainable tourism. For these reasons, the concept of sustainable tourism is fundamental, or must be fundamental, in Brazil as well.

Mielke (2012) has suggested there are three key issues related to the sustainability of community based tourism projects in Brazil. These are access to markets, management of strategic partnerships of interest, and internal governance. Measures of success must be identified early in the planning process, and should be tailored to the specific community in which the project is undertaken.

As a result of these long-term efforts, some success has been seen in the form of sustainable tourism in Brazil, but typically in a micro rather than macro form. Some communities have been involved in efforts where local goods, services and food products have been used as community based tourism products, as we will discuss in the two case studies. These include local guides, lodging and accommodation, handcraft products, and so forth. Community based tourism projects must be carefully thought out, in order to realize the goals of positively impacting the visitor and the local people. Ensuring that community based systems remain sustainable is also a major challenge for non-governmental organizations (NGOs), governmental agencies and other stakeholders.

Educating both tourists and the residents of local communities is a critical variable in maintaining the natural environment in a sustainable manner. Only through education we will be able to achieve sustainable tourism goals. These goals include developing broad awareness and allowing for the understanding of the significant contributions that tourism may bring to the environment and the economy. In addition, one must keep in mind that in order for sustainable tourism to be successful, a strong emphasis on the natural resources is indispensable.

Brazil's Bureau of Tourism suggests that it is necessary to diversify the experiences provided by natural touristic attractions. Accordingly, new ecotourism opportunities, as well as more diverse ecotourism-related services and products must be developed, but in a sustainable manner. If such sustainable development were to occur at a more macro level, the potential of the natural resources could be realized in Brazil. This must be realized, as ecotourism is of such critical importance to the Brazilian government. Brazilian tourism policy has paired ecotourism with a national effort to promote both economic and social development. The unique natural beauty and ecological value of Brazilian landscapes make the country a worldwide destination for ecotourism, albeit there is much room for improvement.

In 2007, the Brazilian government established the Política Nacional de Desenvolvimento Sustentável dos Povos e Comunidades Tradicionais (National Policy of Sustainable Development of Traditional People and Communities or PNPCT). The executive order acknowledged that indigenous and traditional Brazilian people, as well as the culturally diverse communities in which they live, have a strong and important sense of identification. These traditional groups have their own social organization systems and unique ways in which they inhabit and use the land and its natural resources. These include cultural social, religious, ancestral and economic uses, making use of the knowledge, innovations and practices generated and transmitted by tradition (Brasil 2007, Decreto n° 6.040, de 07 de Fevereiro de 2007). There are many such groups in Brazil, such as Indigenous, Maroons, Tappers, Chestnut Collectors, Coco-de-Babaçu Breakers, Faxinalenses, Fishermen, Shellfish Collectors, Riparian, Caissaras, Raftsman, Campieri, Wetland, Caatingueiros among others. It is important to note that the Brazilian government has realized the importance of such traditional groups, and promotes their sustainability.

Another primary objective of the PNPCT was to promote the sustainable development by residents of traditional communities. The executive order provided for a strong emphasis on recognizing, strengthening, and assurance of their territorial, social, environmental, economic and cultural rights in respect and appreciation to their identity, structural systems and institutions.

Any sort of tourism management and/or planning on indigenous lands or in other traditional communities must have its origins in community discussion and participation. Engaging the communities helps to clarify objectives and mitigate doubts and fears often noted by local residents. Using a community based approach also helps NGOs to understand community organizational systems, and considerations needed to ensure that cultural realities are respected. Understanding these particular aspects of sustainable tourism will allow the local residents to make informed decisions.

12.4 Case Studies

In the following case studies, we will discuss two different attempts at sustainable tourism in two distinctly different settings, both located in the southeastern Brazil state of Paraná (Fig. 12.1). Both case studies exemplify the need for an economic benefit to a local community.

The first case study focuses on a unique type of community, Faxinal, and its contribution as a micro level sustainable tourism location. In this case study we discuss how people in this unique community have developed a sustainable lifestyle by producing, marketing and selling organic foods and vegetables, and other products. These are sold at local produce markets as traditional, local products.

Fig. 12.1 Paraná State Map: Prudentópolis and Superagui National Park

We will also discuss the importance of managing the ecosystem and engaging local NGOs in addition to focusing on economic variables.

The second case study involves a small community of residents living on an island located off the coast of southeastern Brazil, nested within the Superagui National Park. This ecotourism setting includes extremely rare species, such as the endangered purple-faced parrot or Papagaio de cara roxa (*Amazona brasiliensis*) and the black-faced lion tamarin (*Leontopithecus caissara*) among others. In the Superagui community, a healthy fishing economy is supplemented by income the provision of services to tourists visiting the island. The economic impact is felt from very small and rustic restaurants, taverns and guest houses rather than from locally produced foods and plants. The local residents were "grandfathered in" when the island was designated as a national park, and no new families are permitted to move to the community to participate in the delivery of tourism services. Both the Faxinais and the Superagui settings are considered "traditional" Brazilian communities, but not indigenous communities.

12.4.1 Case Study 1: Faxinais in Prudentópolis

12.4.1.1 Using Ecotourism as a Tool for Sustainable Development

The first case study focuses on a truly unique traditional community; the Faxinais. The Faxinais are small communities located in southeastern Brazil, focusing on communal production of livestock and vegetables for use in a Brazilian form of localized, garden to table concept. There are 19 Faxinais communities within a single protected area named the Serra da Esperança Area of Environmental Preservation. This case study focuses on one of these, located in Prudentópolis, Paraná,

Some of the last significant remainders of Araucaria Moist Forest are located in Prudentópolis. This city is rich in biodiversity and boasts more than 42 cataloged waterfalls, and most of the area's forested remnants are concentrated in this unique setting. The Faxinais are unique cultural communities that host a similarly unique traditional population. They make up an historical form of social organization of land use whose environmental conditions are considered to be highly preserved in comparison to other communities in the region. Faxinais is a unique style of social organization that can be compared to a communal setting, where forest sites and livestock lands are shared by local families (Chang 1985).

A combination of the unique cooperation within the environment, including both small farm lands and forest settings, with the closely knit social structure allowed for the forests to flourish. The local people worked together to ensure the forests and agriculture lands were maintained in a sustainable manner. The setting is dotted with small, neatly kept homes that often have large, organic gardens surrounding them. This micro level of sustainability, however, has resulted in a setting that is so well maintained that it has become a target of agricultural developers. In fact the forests are so well conserved that they now suffer intense pressure to be converted into large agriculture communal farming plots. Ironically, the Faxinais are in danger because of their sustainability.

The Faxinais are also appealing to tourists for many reasons. They are unique agro-pastoral systems that allow the sustainable existence of nature through practices of family farming and agro-ecology. The forest resources are also used in to grow medicinal plants. The land is divided into planting lands and livestock lands. Some families within the Faxinais are involved in growing and harvesting sustainable natural resources (pinion, mate herb, etc.) that result in a relatively low impact on the environment. The decision making processes about land and crop management, as well as labor issues, are dealt with within community, making it an unusual social setting within Brazil.

According to Oliveira (2008), the traditional productive activities in the Faxinais areas are under permanent and increasing pressure. This pressure has increased as the soybean has become Brazil's primary agricultural crop, which is a direct result of the extremely high earnings per hectare in comparison to other land uses. The influence of the soybean crop has resulted in a tremendous loss of small family farms, as smaller properties were incorporated to larger farms.

Clearly, ecotourism and its services and products have the potential to become a viable alternative to agricultural practices, and can boost the local development of the region. One of the many benefits of ecotourism in the region is that it may reduce a growing problem of rural exodus and foster family agriculture practices. The natural scenic beauty of the Faxinais, combined with the unusual variety of micro-environments result in an ecosystem mosaic that include a unique diversity of natural attractions. Such attractions can result in an increased number of tourist activity, benefiting tourists and locals alike.

In the project entitled "Ecotourism as a tool for sustainable development of Faxinais in Prudentópolis," a non-governmental organization (NGO) offered training courses that approached issues such as sustainable tourism, hospitality, biodynamic agriculture, business management, and other topics. Those courses were sponsored by the Brazilian Ministry of Environment and supervised by the NGO Institute Guardians of Nature. The objective of this project was to select families that had the interest and the potential to host tourists, and facilitate their ecotourism development efforts.

Since 2007 the NGO has been conducting market surveys that assisted in the identification of trails, environmental diagnosis and geo-referencing of Faxinais at the selected properties. The surveys are used in the development of business plans and to study the potential of some key ecotourism attractions. The primary objectives of the study were to incorporate a visitor use monitoring plan, and to assist in the marketing and the design of a ubiquitous Faxinais logo (A logo with pictures of local flora and fauna livestock, a pine tree, a wagon and riparian forest next to a river) was created in order to convey the image of the Faxinais) (Fig. 12.2).

The results of the survey showed that by empowering the community to work within the ecotourism framework, a first step toward transforming the Faxinais into an economic model of sustainable development was taken. This was done by providing for a better quality of life through economic performance and demonstrating the importance of community-based conservation of the Araucaria Forest to the local inhabitants. The final result was that specific strategies were implemented to develop and market the tourism route, including brochures and a website. Additionally, a partnership with official tourism offices in national and international fairs, focusing on tourism promotion, was developed to stir interest.

Fig. 12.2 Logo for the Faxinais route

12.4.2 Case Study 2: Superagui National Park and the Community

Superagui National Park is located in the city of Guaraqueçaba, on an island along the north coast in the state of Paraná, in the south of Brazil. The park is accessible only by boat trip (1 h via rapid boat, 3 h via traditional boat) from the gateway city of Paranaguá.

Superagui Island has had a long history of recognition as a natural resource setting of great value (Fig. 12.3). In 1970, Superagui Island was considered listed as a natural area by the State of Paraná. In 1991, shortly after being named a National Park, Superagui National Park was included within the Biosphere Reserve "Vale do Ribeira – Serra da Graciosa/Mata Atlântica" by the United Nations Educational, Scientific and Cultural Organization (UNESCO). In 1998 the region was declared a Natural Heritage of Humanity setting within the "Floresta Atlântica: Reservas do Sudeste" site. Finally, in 2012, the major cultural event in the region, Fandango, was registered by the Institute of National Historical and Artistic Heritage (IPHAN) as a cultural good, known as Intangible Heritage (IPHAN 2013).

The population and its surroundings are considered *traditional* (but not *indigenous*), made up primarily by artisanal fishermen. The setting is noted as one the most significant continuous remnant of rainforest in Brazil. This designation was awarded primarily because of the amount of biological and landscape diversity in its beaches, mangrove areas and sandbanks, as well as its numerous rivers and surrounding bays (Fig. 12.4).

There are three dimensions of sustainability that best express the relationship of ecotourism within the Superagui National Park; environmental, social, and economic. Regarding the environment, biodiversity conservation has a high degree

Fig. 12.3 Superagui Island, at Superagui National Park

Fig. 12.4 Praia Deserta, at Superagui National Park

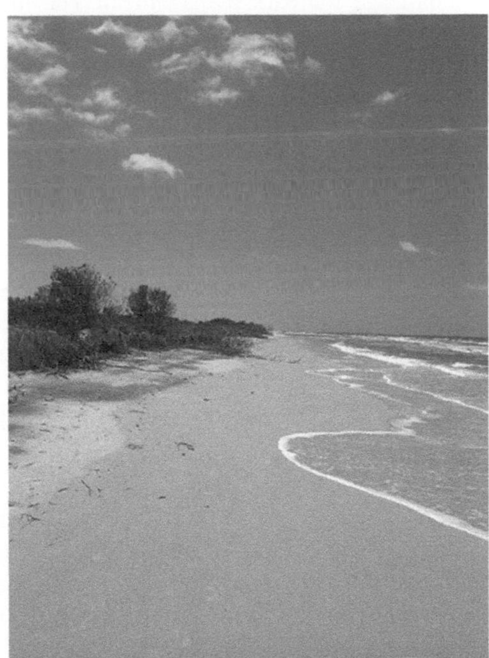

of importance in the Superagui National Park. As noted previously, the park provides refuge for endangered species such as the papagaio-da-cara-roxa (*Amazona brasiliensis)*, the black-faced lion Tamarin (*Leontopithecus caissara*), dolphins such as the boto-cinza *(Sotalia guianensis*), the jacaré do papo amarelo (*Caiman latirostris)*, and the porpoise (*Pontoporia blainville*), which is the most endangered cetacean in the South Atlantic Ocean. The island is also home to migratory birds and the guará (*Eudocimus ruber*), a red heron that virtually disappeared from the area for years but has been reappearing over the past several years. The ecosystem of mangroves and sandbanks include key species in the Atlantic forest conservation such as the jussara palmetto (*Euterpe edulis*), the guanandi (*Calophyllum brasiliense*) and the jerivá (*Syagrus romanzoffiana*), which serves as food source for monkeys (ICMBio 2014).

Tourism will most likely not cause a direct negative impact on these natural resources so long as the number of tourists remains low. According to information provided by Superagui National Park management, approximately 4500 visitors stay in the 26 guesthouses or at campsites in the park region every year, with most visits taking place during the Carnival and New Year celebrations. The vast majority of visitation by tourists occurs during the summer, when boat access becomes easier. Boat travel is limited during the winter months, when the open water may be too rough for tourists. Aside from the logistics of travel to the island, there is a limited supply of potable drinking water on the island, and sewage disposal will remain a challenge for park management. The local community places a great deal of emphasis on waste collection, with recycling a large part in the effort.

Focusing on the social component, 14 small communities are located in the boundaries of the Park, totaling approximately 1600 residents. The region's population fluctuates significantly as a result of the practice of nomadism. One community has only one resident, another community's population ranges between three and 280 inhabitants, and the largest has 780 residents.

The dominant discourse concerning tourism on the island is over the constraints imposed by the creation of the National Park and the changed relationship between the residents and the natural resources. A social carrying capacity number had been determined prior to the area being named a park, which allowed for a sufficient number of visitors to make an economic impact on the community. Without the park visitors, the people living in the local communities would have virtually no income outside of a meager income from fishing and shrimping. The guest houses, or "pousadas" that are available to tourists are very simple, with small rustic "hostel-like" amenities. Simplicity is also seen in the island's cuisine, as all foods are made locally and served by a local resident. No motorized vehicles exist on the island, although small motorized boats are used to transport local residents and tourists to and from various places.

In traditional terminology, the Brazilian legislation refers to the locals as caissaras fishermen. The caissaras are traditional anglers and boaters; iconic watermen, immersed in traditional maritime technology (Adams 2000). These fishermen have evolved from traditional family farmers to fishermen over time, as the opportunity to make a living from fishing evolved into a better way of life for them and their families. This evolution occurred primarily between 1930 and 1950, and the change resulted in the near extinction of traditional farming on the island. The caissaras fishermen are a large part of the cultural fabric of the island, and are a cultural attraction to tourists. The men are sources of information for tourists, as their experiences can be easily shared with the tourists along the sandy beaches. Aside from being artisanal anglers, the fishermen also know and practice oyster farming, make handicrafts from bones and shells, and other traditional practices. These traditional fishermen also practice the art of folklore, telling stories about traditional island fishing traditions and superstitions. They dabble in the use of medicinal herbs, and generally provide a much appreciated, authentic traditional experience for tourists.

One of the most important cultural elements of this population is the traditional Fandango, which is considered as an "intangible Brazilian heritage" by the Brazil federal government. This cultural practice has been characterized as Fandango caissara, and is a choreographic-musical-poetic and festive expression. In essence, fandango is festive traditional folk music, played on handmade instruments. Fandango occurs nightly in local communities throughout the southern coast of Brazil and extends northward into the states of São Paulo and to the north coast of the state of Paraná.

The influx of tourism has added value to the unique characteristics of this art form. The fandango experience is highly sought by visitors for both leisure experiences such as learning how to play an instrument. One of the most popular fandango instruments is the "rabeca," a kind of violin made with local wood and played much like a fiddle in the style of Appalachian bluegrass), and other cultural aspects. However, there has been a backlash from local religious groups who forbid their group members from participating in the art. It is seen by some as simply "partying" and alcohol consumption, which may inhibit the group's role of healing and saving the souls of local residents.

Perhaps the economic variable is the most important topic to discuss within the context of this chapter. Even with restrictions on the use of natural resources by locals, artisanal fishing within the communities is encouraged, as it is necessary for family survival. The expansion of tourism on the island has provided local residents with an alternative to their life at sea; perhaps an evolution from artisanal fishing to sustainable tourism services. These include food, transportation, the making and selling of local handicrafts, among others.

One prime example of a community-based economic activity is the manufacturing of Cataia, an alcoholic drink (with cachaça, a sugar cane-based liquor), also known as "caissara whiskey." The Cataia was established by a Women's Association from Barra do Ararapira, with the support of park management, from a locally grown plant named pepper pseudocaryophyllus. The cachaça is marinated in the plant leaves and its consumption is enjoyed by visitors and locals alike. Many tourists are enchanted by this locally produced beverage specifically because the plant leaves are harvested only in a specific small area of the island.

12.5 Conclusions

The two case studies in this chapter demonstrate two distinctly different strategies, in different settings, that have resulted in what can be defined as sustainable tourism. Referring to Table 12.1, both studies include, to a certain degree, the components that are seen in the sampling of previous case studies around the world.

People living within the local community have been positively impacted through the economic impact, governmental agencies and non-governmental organizations were actively involved in the efforts, and the protected environment in which the communities are nested played a large role.

These successes are the result of an effort by the Brazilian government, both federal and state, to invest in sustainable tourism at a community level. For over two decades, Brazil has been developing sustainable methods of tourism that are designed to find the delicate balance between tourism and local impacts. These case studies were developed as a direct result of governmental policy, non-governmental organizations were involved in the process, and the local community was engaged in the process at the outset. The projects were designed to protect the flora and fauna while celebrating the rich cultural resources that overly the natural resources. The social component is an important variable in that both the local communities and the tourists must perceive a positive impact. In almost all cases of successful sustainable, community-based tourism, this delicate balance must be struck and maintained within the community.

For the Faxinais, the project has shown that it is possible to assist residents in generating income, while still promoting the conservation of the land and offering sustainable alternatives to these communities. If it were not for the economic impact, it is doubtful that the Faxinais would have been able to even continue living in the community.

In discussing solutions, we often refer to tools that can be used by local communities, governments, and non-governmental entities. We suggest it is necessary to encourage the establishment of future programs and implement projects that empower communities to increase their sustainability. The Faxinais and the residents of Superagui, as well as other Brazilian traditional communities, are links in the chain of environmental, cultural and self-sustainable model of community preservation. Many Brazilian communities are conscious about the importance of their organization for the achievement of sustainable development in their community and to conserve natural resources and cultural heritage. These two case studies, though presented in a micro lens, are representative of other successful case studies that have shown common models of success, or tools, in developing sustainable ecotourism.

At the community level, some specific tools or solutions can be suggested, such as focusing on customer satisfaction, marketing efforts, and understanding the various segments of potential tourists (Ahmed 2010). By designing tourism packages that focus on likely potential users, the community members may be more successful in developing sustainable tourism efforts.

Finally, it is important to stress the importance of following the prescriptive methods outlined in the vast literature on the topic of ecotourism and sustainability. At a macro level, it is clear that we still struggle to answer the basic questions posed by Wall (1997) (*is ecotourism sustainable?*), and by Weaver (2006) (*is the concept of sustainability so utopian that it may be impossible to achieve?*). However, at a micro level, within specific communities, and within specific nations around the world, small successes are frequently seen. As we continue to study what makes a successful, sustainable ecotourism program in a community, perhaps we will begin to see threads and trends that will lead us to answer these very important and very complex questions.

References

Adams, C. 2000. As populações caiçaras e o mito do bom selvagem: A necessidade de uma nova abordagem interdisciplinar. *Revista de Antropologia* 43(1): 145–183.

Ahmed, F. 2010. Factors affecting the selection of tour destination in Bangladesh: An empirical analysis. *International Journal of Business and Management* 5(3): 52–61.

Alaeddinoglu, F., and A.S. Can. 2011. Identification and classification of nature-based tourism resources: Western Lake Van basin, Turkey. *Procedia Social and Behavioral Sciences* 19: 198–207.

Alam, M., Y. Furukawa, and S. Akter. 2009. Forest-based tourism in Bangladesh: Status, problems and prospects. *Tourismos: An International Multidisciplinary Journal of Tourism* 5(1): 163–172.

Araujo, L.M., and D. Dredge. 2012. Tourism Developemt, Policy and Planning in Brazil. In *Tourism in Brazil: Environment, management and segments*, ed. G. Lohmann and D. Dredge, 17–31. London: Routledge.

Blangy, S., and M.E. Wood. 1993. Developing and implementing ecotourism guidelines for wildlands and neighbouring communities. In *Ecotourism, a guide for planners and managers*, vol. 1, ed. K. Lindberg and D.E. Hawkins, 32–54. Burlington. The Ecotourism Society.

Boyd, S.W., and R.W. Butler. 1996. Managing ecotourism: An opportunity spectrum approach. *Tourism Management* 17(8): 557–566.

Brasil. 2007. Decreto No 6.040, de 7 de Fevereiro de 2007. http://www.planalto.gov.br/ccivil_03/_Ato2007-2010/2007/Decreto/D6040.htm. Accessed 7 May 2014.

Buultjens, J., M. Tiyce, and D. Gale. 2003. Sustainable forest-based tourism in Northwest New South Wales, Australia: A problematic goal. *Tourism Review International* 7(1): 1–12.

Campbell, L.M. 1999. Ecotourism in rural developing communities. *Annals of Tourism Research* 26(3): 534–553.

Chang, M.Y. 1985. *Sistema Faxinal: uma forma de organização camponesa em desagregação no centro sul do PR*. A thesis submitted in partial fulfilment of the requirements of Rio de Janeiro Federal University for the Degree of Master. Rio de Janeiro: Rio de Janeiro Federal University.

Gouvea, R. 2004. Managing the ecotourism industry in Latin America: Challenges and opportunities. *Problems and Perspectives in Management* 2: 1–9.

Hill, J.L., and T. Gale (eds.). 2009. *Ecotourism and environmental sustainability: Principles and practice*. Farnham: Ashgate.

ICMBio – Instituto Chico Mendes de Conservação Da Biodiversidade. Parque Nacional do Superagui. 2014. http://www.icmbio.gov.br/portal/o-que-fazemos/visitacao/ucs-abertas-a-visitacao/209-parque-nacional-do-superagui. Accessed 11 Jun 2013.

Iphan. 2013. *Dossiê de registro do Fandango Caiçara*. Paranaguá-PR: Associação Cultural Caburé.

Lai, P.H., and S.K. Nepal. 2006. Local perspectives of ecotourism development in Tawushan Nature Reserve, Taiwan. *Tourism Management* 27: 1117–1129.
Mielke, E.J.C. 2012. Community-based tourism. In *Tourism in Brazil: Environment, management and segments*, ed. G. Lohmann and D. Dredge, 30–43. London: Routledge.
Navalporto, J.A.S., F.G. Quiroga, and M.S. Perez. 2012. Evaluation of tourism development in the National Parks of Spain. *International Journal of Business and Social Science* 3(14).
Oliveira, D.A. 2008. *Os Faxinais do Município de Prudentópolis- PR: Potencialidades e Perspectivas para o Turismo Rural.* A thesis submitted in partial fulfilment of the requirements of Univali University for the Degree of Master of Tourism. Camboriú: Univali University.
Palhares, G.L. 2012. *Tourism in Brazil: Environment, management and segments*, vol. 3. London: Routledge.
Reimer, J.K., and P. Walter. 2013. How do you know it when you see it? Community-based ecotourism in the Cardamom Mountains of southwestern Cambodia. *Tourism Management* 34: 122–132.
Ross, S., and G. Wall. 1999. Ecotourism: Towards congruence between theory and practice. *Tourism Management* 20: 123–132.
Silva, G., and M.E. McDill. 2004. Barriers to ecotourism supplier success: A comparison of agency and business perspectives. *Journal of Sustainable Tourism* 12(4): 289–305.
Sisto, P.Z. 2003. *Turismo sustentable: Es possible en Argentina?* Buenos Aires: Ediciones Turisticas.
Torres-Sovero, C., J.A. Gonzalez, B. Martin-Lopez, and C.A. Kirkby. 2012. Social-ecological factors influencing tourist satisfaction in three ecotourism lodges in the southeastern Peruvian Amazon. *Tourism Management* 33: 545–552.
Tsaur, S.H., Y.C. Lin, and J.H. Lin. 2006. Evaluating ecotourism sustainability from the integrated perspective of resource, community and tourism. *Tourism Management* 27: 640–653.
Wall, G. 1997. Forum: Is ecotourism sustainable? *Environmental Management* 21(4): 483–491.
Weaver, D.B. 2006. *Sustainable tourism: Theory and practice*. London: Routledge.
Wright, P.A. 1995. Sustainable ecotourism: Balancing economic, environmental and social goals within an ethical framework. *Tourism Recreation Research* 20(1): 5–13.
Wunder, S. 2000. Ecotourism and economic incentives – An empirical approach. *Ecological Economics* 32: 465–479.

Chapter 13
Tourism and Social Capital: Case Studies from Rural Nepal

Martina Shakya

Abstract Tourism has a wide range of impacts on developing societies. The positive impacts of tourism on local economic growth are widely acknowledged. Concern has been voiced, however, about the social and cultural impacts of tourism due to observed changes in local norms, values and behaviour. This chapter proposes the concept of social capital to analyze the social and cultural implications of tourism in Nepal. Empirical evidence from a household survey and village case studies reveals a decline of bonding social capital and an increase in bridging social capital due to tourism. Tourism can exacerbate local conflicts and reduce the relevance of indigenous self-help mechanisms. At the same time, the formation of new institutions and opportunities to develop and expand hierarchical, extra-community networks have been promoted by tourism. Highlighting the interdependencies and trade-offs between economic advancement and changes in social capital for sustainable development, the chapter calls for a more pragmatic debate on tourism's social impacts in developing countries.

Keywords Social capital • Tourism • Poverty • Risk • Rural development • Nepal

13.1 Introduction

Tourism has a wide range of impacts on the economy, the environment and the people living in destinations (Wall and Mathieson 2006; Harrison 1992). Undeniably, tourism also has an influence on society and culture, particularly in rural communities of developing countries, by changing family relations, the economic status of individuals and the meaning of local institutions and networks (Harrison 1992). This chapter proposes the concept of social capital as a framework to analyze tourism's social impacts and to assess the consequences of such impacts

M. Shakya (✉)
Institute of Development Research and Development Policy, Ruhr University Bochum, Universitätsstr. 150, Bochum 44801, Germany
e-mail: martina.shakya@rub.de

S.F. McCool, K. Bosak (eds.), *Reframing Sustainable Tourism*,
Environmental Challenges and Solutions 2, DOI 10.1007/978-94-017-7209-9_13

for destination communities. Briefly defined as "the norms and networks that enable people to act collectively" (Woolcock and Narayan 2000, p. 226), social capital is regarded as a source of human welfare, complementing conventional asset categories such as natural, physical and human capital (Grootaert 1998).

Based on evidence from four rural communities of Nepal, one of the poorest countries in the world, we scrutinize whether and how tourism influences social capital and sustainable development in destinations. After a literature review, the concept of social capital is elaborated as a research framework for this study. The geographical setting of the study and the research methodology are then described. Drawing on evidence from household surveys and village case studies, the empirical results are presented in detail. The chapter ends with conclusions and lessons learnt.

13.2 Social and Cultural Impacts of Tourism

As one of the most dynamic economic activities and the world's largest generator of wealth and jobs, tourism has been hailed as a pathway to prosperity for poor, developing nations (UNWTO 2002; WTTC 2007; Mitchell and Ashley 2007, 2010). Through the creation of income and employment, tourism can contribute to poverty alleviation and economic growth, especially in rural peripheries such as coastal, mountain and desert areas, thereby reducing regional imbalances (Wiggins and Proctor 2001; Ashley and Maxwell 2001). Whereas the economic impacts of tourism are mostly regarded as desirable, it is more difficult to arrive at general judgments of the environmental and social effects of tourism (Harrison 1992; Vorlaufer 1996; Mihalič 2002; Telfer and Sharpley 2008). The expansion of tourism into ecologically fragile areas can lead to local environmental problems but can also provide incentives for sustainable resource use and conservation (Southgate and Sharpley 2002; Wall and Mathieson 2006; Boo 1990; Wells et al. 1992; Telfer 2002).

Even more difficult to judge are the social and cultural implications of tourism. Tourism has been accused of promoting inequality and social tension, changing local power structures and "crowding out" traditional occupations such as farming (Turner and Ash 1975; De Kadt 1979; Bachmann 1988; Graburn and Jafari 1991). Previously less influential or discriminated social groups, including women and ethnic minorities, may gain status and economic strength as a result of tourism (Harrison 1992; Coppock 1978; Nepal 2005; Gurung 1995). However, people's social position might also deteriorate, if they get involved in low-status jobs or exploitative forms of tourism (Cohen 1988). Tourism may perpetuate existing inequality, if indigenous entrepreneurs and tourism employees emerge from the already wealthy and influential segments of rural society (Nepal 2005). If outside investors move in and take ownership of the local tourism industry, this may result in a division between local (small-scale) and "metropolitan" (large-scale) interests.

Some authors have suggested that the commoditization (or commercialization) of local cultures changes the "meaning of cultural products and human relations," leading, ultimately, to a loss of cultural identity (Cohen 1988, p. 372). This is exemplified by indigenous art production moving from "functional traditional art" towards commercial production of souvenirs for the tourist market (Harrison 1992; Vorlaufer 1996; Hashimoto 2002). On the other hand, tourists' interest in local cultures has been found to strengthen or revive cultural practices and art forms, reinforcing cultural pride or even leading to an invention of new cultural institutions (Harrison 1992; Vorlaufer 1996). The personal interaction between affluent tourists and poor host societies has been described as an asymmetrical relationship due to the cultural and economic distance between the two parties. The adoption of "Western" values and patterns of behavior by members of the host society has become known as the demonstration effects of tourism (Harrison 1992; Hashimoto 2002).

Implicit in many studies on the social and cultural effects of tourism is the evaluation of such impacts as negative and, hence, undesirable. Without doubt, tourism changes the local society, culture and economy, with potential negative effects such as crime, prostitution or conflict. However, the explicit or implicit view that the social and cultural features of poor societies are "weak and in dire need of protection from outside" is equally contentious:

> There is no inherent virtue, for example, in the extended family, which may as often be a source of repression and autocratic (and patriarchal) control as one of security and freedom. Similarly, the superiority of palm wine over Western beers, or of traditional dress over blue jeans, may be affirmed but, ultimately, is a matter of taste. (Harrison 1992, pp. 30–31)

In conclusion, the social and cultural impacts of tourism on rural societies defy any universal judgment and often appear tainted by researchers' subjectivity. Consequently, a more pragmatic and less normative debate on the social consequences of tourism is desirable. The concept of social capital, as further elaborated in the following section, is proposed here as a research framework to analyze the social and cultural impacts of tourism.

13.3 What Is Social Capital?

Researchers from various academic backgrounds have noticed that regional and cross-country differences in economic performance could not be explained exclusively by economic variables. It was found that economic action in society was "embedded in structures of social relations" and could not be separated from these (Granovetter 1985, p. 481). Personal relationships, networks, associations, institutions, norms and values influence the economic success of a region or country by promoting trust, information-sharing, political power and cooperative action. Over time, scholars have suggested various definitions and conceptual frameworks to analyze what we now popularly refer to as *social capital*, i.e. "the institutions, the relationships, the attitudes and values that govern interactions among people and contribute to economic and social development" (Grootaert and Van Bastelaar 2001, p. 4; cf. Bourdieu 1986; Coleman 1988; Putnam 1993; Feldman and Assaf 1999). Since the 1990s, the concept of social capital has entered the mainstream of development research and development policy.

Like other forms of capital, social capital is no natural given. It must be acquired through investment strategies oriented to the institutionalization of group relations and can be accumulated or lost (Portes 1998; Ostrom 2000; Grootaert and Van Bastelaar 2001, cf. Scoones 1998). Social assets—e.g. claims on family and community members, mutual support mechanisms and social networks—can provide informal insurance in situations where formal insurance arrangements are absent or inaccessible, as common in many parts of the developing world (Swift 1989). Unlike other assets, social capital can only be held by a group of people and requires some cooperation among its members (Grootaert 1998).

An analytical distinction can be made between the scope, forms and channels of social capital (Table 13.1). As regards *scope*, social capital can be analyzed at the micro, meso and macro levels. Micro- and meso-level social capital have also been described as "bonding" and "bridging" social capital (Woolcock and

Table 13.1 Typology of social capital

Dimension of social capital		Characteristics	Manifestations
Scope	Micro-level, "bonding"	Horizontal, interpersonal (intra-group), homogeneous groups, mostly informal	Local associations, networks, interest groups; associated norms and values
	Meso-level, "bridging"	Vertical, inter-group, hierarchical, heterogeneous groups, informal or formal	Regional associations/networks of personal and/or corporate actors (mostly groups)
	Macro-level	Social and political environment, most formalized institutional relationships	Political regime, rule of law, court system, civil and political liberties
Form	Structural	Relatively objective, externally observable, informal or formal	Networks and other social structures, supplemented by rules, procedures and precedents
	Cognitive	Subjective, intangible, informal	Shared norms, values, trust, attitudes, beliefs
Channel	e.g. Cooperation Information-sharing Collective action	Reciprocal or unidirectional; informal or formal; reduction of transaction costs (information costs, negotiation costs, enforcement costs)	Networks, associations, copying

Source: Own compilation, based on Grootaert and Van Bastelaar 2001, pp. 4–6

Narayan 2000; Putnam 1993, 1995; Coleman 1990). *Bonding social capital* refers to networks of homogeneous, intra-community groups with a common interest and strong social cohesion. Correspondingly, vertical institutions, i.e. extra-community networks have been described as *bridging social capital*. Empirical studies suggest that bridging social capital has positive effects on growth, access to markets and upward economic mobility of rural households. In contrast, bonding social capital has been linked to poor and destitute households that are just "getting by" (Beugelsdijk and Smulders 2003; Woolcock and Narayan 2000; Esman and Uphoff 1984). Finally, the *macro perspective* looks at the "social and political environment that shapes social structure and enables norms to develop"—with a critical effect on the rate and pattern of economic development (Grootaert and Van Bastelaar 2001, p. 5, cf. Olson 1982; North 1990). Micro, meso and macro level social capital are coexistent and work in complementary ways. Moreover, some substitution between different forms of social capital is possible (Grootaert and Van Bastelaar 2001).

Another distinction can be made between two *forms* of social capital. *Structural social capital* refers to tangible, observable manifestations. It is more difficult to analyze intangible *cognitive social capital*, such as norms, values and trust. Finally, social capital can be analyzed by the *channels* through which it impacts upon a society's development. For instance, transmission of knowledge can be facilitated by trust and information-pooling within horizontal associations, resulting in reduced transaction costs and income increases (Grootaert and Van Bastelaar 2001; Ostrom 1990; Narayan and Pritchett 1997).

The virtue of social capital as a conceptual framework for development studies is obvious, considering the range of possible applications (cf. Ostrom 1990; Hirschman 1984; Esman and Uphoff 1984; Uphoff 1993, 2000, Woolcock and Narayan 2000). But authors have also noted some negative effects of social capital. Extreme examples of a "dark side" of social capital are criminal gangs, cartels and mafia organization (Ostrom 2000, cf. Olson 1982; North 1990) Woolcock (1998, p. 186) therefore concludes that social capital is a "crucial, but enigmatic, component of the development equation," precisely because it can enhance, maintain, or destroy other types of capitals.

To date, only a few authors have explicitly or implicitly linked the concept of social capital with tourism. It has been observed that the causality between tourism and social capital can work in both directions (Macbeth et al. 2004). Some authors take the view that the existence of social capital within a destination community can promote the sustainability of tourism development (McGehee et al. 2010; Claiborne 2010). However, corresponding to the discussion of social and cultural impacts in the previous section, we assume the reverse causality and hypothesize that *tourism has an effect on the social capital of poor, rural societies*, with positive or negative implications for sustainable development. Exploring such impacts in rural communities of Nepal, the empirical analysis will focus on structural manifestations of bonding and bridging social capital.

13.4 Methodology

13.4.1 Case Study Setting

The Federal Democratic Republic of Nepal, a least developed country with considerable tourism potential, was chosen as the geographical setting of the study. With 95 % of the poor living in rural areas, poverty in Nepal is primarily a rural phenomenon (CBS et al. 2005). Poverty not only differs between rural and urban areas; it is considerably lower in the narrow lowland stretch in the South as compared to the Nepalese hills and mountain belts, indicating a "geographical disadvantage" of extremely remote and isolated locations (World Bank 2006). With an estimated 8.2 % share of national GDP and 7 % of total employment, the Nepalese travel & tourism economy has a modest macroeconomic importance (WTTC 2014). Tourism plays a more significant role in the local economy of Nepal's rural destinations, where trekking tourism, mountaineering and wildlife excursions take place. The Himalayan ranges and the wilderness areas of Nepal's lowland belt are important assets of the Nepalese tourism industry. This is exemplified by Langtang National Park and Chitwan National Park, two of Nepal's major tourist destinations. Rasuwa district, which includes the dominant share of Langtang National Park, is one of the most important trekking and mountaineering destinations in Nepal, whereas tourist activities in Chitwan district focus on nature-based activities such as jungle safaris and bird watching in the national park. These two districts were selected for the empirical investigation, as they represent different topographical, ecological and cultural zones of Nepal.

13.4.2 Research Design

To identify impacts of tourism on social capital, a tourism village and a "matching" non-tourism village were identified in each district. The non-tourism villages, Gatlang (Rasuwa) and Shaktikhor (Chitwan) were selected for their structural similarity with the respective tourism village as regards their tourism potential, topographical setting, ecological zone, accessibility, poverty prevalence and ethnic composition. With differing shares of tourism households thus being the main distinguishing variable, the selection of tourism and non-tourism villages in both districts was suitable to detect impacts of tourism on social capital (Table 13.2).

The research methodology combines the rigor of quantitative analysis with explanatory insights from qualitative research. Based on a comprehensive questionnaire, standardized interviews were conducted with 259 households from the four villages and the data entered into an SPSS database. A "treatment group" of *tourism households* (defined as being economically involved in and earning at least part of their income from tourism) could thus be compared with non-tourism households to detect causal links between tourism and social capital. The selection

Table 13.2 Description of case study communities

	Rasuwa district (mountains)		Chitwan district (lowlands)	
Village	Thulo Syabru	Gatlang	Sauraha	Shaktikhor
Group	Treatment group	Comparison group	Treatment group	Comparison group
Total households	122	223	231	183
Population	520	1,183	1,107	829
Sample size (no. of households)	51	70	77	61
Households covered by survey	42 %	31 %	33 %	33 %
Altitude	2210 m a.s.l.	2238 m a.s.l.	250 m a.s.l.	355 m a.s.l.
Topographical and ecological zone	Midhills (mountainous, fans/slopes)	Midhills (mountainous, fans/slopes)	Lowlands (alluvial plains)	Lowlands (alluvial plains to soft slopes)
Distance from highway	2 h walk to district road (unpaved)	4 h walk (2 h drive) to district road (unpaved)	6 km (25 min drive) to national highway (paved)	15 km (40 min drive) to national highway (paved)
Located on road	No (nearest road 2 h away)	Yes (unpaved)	Yes (partly paved)	Yes (partly paved)
Ethnic composition	100 % Tamang	99 % Tamang	Mixed	Mixed
Employment in tourism	14 % of population	<1 % of population	36 % of households	2 % of population

Data sources: MCTCA et al. 2005, 2006; NTB and TRPAP 2003; Bacchyauli VDC 2006; own surveys

Table 13.3 Sample composition

Total sample: 259 households (100 %), thereof			
Tourism households (treatment group)	107 (41 %)	Non-tourism households (control group)	152 (58 %)
Households in tourism villages	128 (49 %)	Households in non-tourism villages	131 (51 %)
Households in Rasuwa district (mountains)	121 (47 %)	Households in Chitwan district (lowlands)	138 (53 %)

of two geographically distinct districts for the empirical field study aims at revealing contextual dimensions of social capital (Table 13.3).

The village case studies aimed at qualitative, explanatory insights about tourism's impact on social capital and embedding theoretical constructs into their real-life context. In addition to the household-level data, community meetings, semi-structured focus group discussions and in-depth interviews with key informants

Table 13.4 Research variables and associated indicators

Research variables	Indicator
Dependent variables	
Bonding social capital	Membership in formal institutions: yes/no (dummy)
	Number of household members involved in these institutions
	Existence of informal institutions/self-help mechanisms
Bridging social capital	Extra-community relationships with relatives/family members
	In the capital Kathmandu
	In another urban area of Nepal
	In a foreign country
Independent variable	
Tourism	Household is economically involved in tourism: yes/no (dummy)
	Tourism income (absolute; percentage share)
	Household is located in a tourism village: yes/no (dummy)
	Tourism income (absolute; percentage share)
Control variable	
Geographical location	Household is located in the mountains (Rasuwa district) or in the lowlands (Chitwan district) (dummy)

were held in each case study village. The two methodological pillars—statistical analysis of the aggregate household data plus village case studies based on the partial samples, group exercises and in-depth interviews—were expected to provide complementary and mutually enriching information with regard to tourism's impact on social capital in the four villages (cf. Shakya 2009).

Based on the social capital typology elaborated in the previous section (cf. Table 13.1), we will explore empirically observable, structural expressions of bonding and bridging social capital and examine their connection with tourism (Table 13.4). Households' membership in formal, horizontal institutions such as community-based organizations and functional groups is assessed as a proxy for bonding social capital. As proxies for bridging social capital, we will examine extra-community, vertical networks by exploring whether households have relatives or family members in Kathmandu, in other urban areas of Nepal or in foreign countries. We also look at multi-locality of households, assuming that migrant family members contribute to the formation of extra-community networks in the same manner than relatives or family members permanently living in another location.

13.5 Impacts of Tourism on Social Capital: Quantitative Evidence

We begin our empirical exploration by analyzing the aggregate household data. Regression analysis (significance level: $p \leq 0.05$) reveals whether the selected indicators of social capital are systematically dependent on households' economic

Table 13.5 Multiple regression analysis (OLS) of the impact of tourism and geographical location on social capital (n = 259)

Independent variables:	Dependent variable				
	SC 1	SC 2	SC 3	SC 4	SC 5
Tourism (0 = non-tourism HH, 1 = tourism HH); $beta_1$ (standardized)	–	0.154*	0.130*	–	–
Geography (0 = Rasuwa/mountains; 1 = Chitwan/lowlands); $beta_2$ (standardized)	0.157*	0.253**	–	0.350**	0.210**
R	0.159*	0.318**	0.134	0.355**	0.207*
R^2 (adjusted)	0.018	0.094	0.010	0.119	0.035

Data source: own survey
Significance (2-tailed): **p = 0.01, *p = 0.05 (only significant coefficients are shown)
Dependent variables:
SC 1: Household is involved in any formal institution(s) (dummy)
SC 2: No. of household members who are involved in formal institution(s)
SC 3: Household has family/relatives in Kathmandu (dummy)
SC 4: Household has family/relatives in an urban area (dummy)
SC 5: Household has family/relatives in a foreign country (dummy)

involvement in tourism. We also wanted to test whether social capital is influenced by contextual variables, comparing household data from the mountains and the lowlands.

Table 13.5 presents the results of the regression analysis (OLS) with regard to the defined indicators of social capital. Controlling for the impact of lowland residence, tourism involvement is not associated with households' membership in *formal institutions* per se, but has a positive impact on the number of household members participating in such institutions (cf. variables SC 1 and SC 2). The indicator also suggests that tourism households are more likely to be involved in more than one institution than non-tourism households. This appears plausible; most tourism households are members of tourism organizations—which exist even in the non-tourism communities—in addition to their membership in non-tourism organizations. Such "institutional diversification" corresponds with the greater livelihood diversification of the tourism households, as the overwhelming majority of households pursue tourism as an additional income source, along with farming and other activities.

In contrast, the empirical data do not reveal an impact of tourism on *bridging social capital* in the form of extra-community networks (cf. variables SC 3–5). Controlling for the impact of lowland residence, the partial coefficients for extra-community relations are insignificant. Concluding thus far, no statistically significant difference in social capital could be found between tourism and non-tourism households across the four Nepalese villages. Instead, the analysis indicates that households in the lowland district of Chitwan tend to be more frequently involved in formal institutions and extra-community networks than households in the mountainous district of Rasuwa. This may be explained by the better accessibility of rural areas in the Nepalese lowlands, which facilitates social and economic transactions. To identify location-specific differences with regard to tourism's impact on social capital and to cross-validate the quantitative findings, we now turn to the village case studies.

13.6 Tourism and Social Capital in the Nepalese Mountains

13.6.1 Thulo Syabru, a Tourism Community in Rasuwa District

Thulo Syabru is an old settlement almost exclusively inhabited by one ethnic group, the Tamang. The village area is located inside Langtang National Park but designated as part of the park's buffer zone. The economic mainstay of households in Thulo Syabru is mixed farming, i.e. a combination of agriculture and animal husbandry. Involving 59 % of households, tourism comes second after farming in the local economy.

Tourism in Thulo Syabru started about 35 years ago, soon after the establishment of Langtang National Park in 1976. Apart from the strategic location on one of Nepal's most popular trekking routes, the scenic setting of the village on a mountain ridge and the Tamang culture are features that attract foreign trekking tourists. The "modern" appearance of many dwellings, most notably the trekking lodges with their cemented walls and corrugated iron roofs, is in stark contrast to the few remaining Tamang houses, which are built of stone and wood. Despite some recent tourism decline, it is believed that a major portion of the 4500 international tourists that visited Langtang National Park in 2006 also stayed at Thulo Syabru (cf. MCTCA 2006). Tourists are accommodated in one of the 14 trekking lodges or on private campsites. All lodges are run by local residents and almost exclusively employ family labor. Facilities range from simple lodges to hotel-type accommodations with up to 23 beds, private bathrooms and solar-heated showers. Additional income opportunities from tourism include seasonal employment as a guide or porter and selling farm products and handicrafts to the trekking tourists. Foreign tourists typically spend one night in Thulo Syabru, using the village as an overnight stop during a trekking tour to Langtang valley or Gosainkunda lake.

Almost three quarters (71 %) of households in the Thulo Syabru sample are a member of one or more of the 12 *formal institutions* that exist in the village. Three village-level organizations are related to tourism or have been established for tourism purposes. A "lodge management committee" had been formed in the 1990s to standardize rates for food and accommodation. As reported, members had defaulted on the group's rules by undercutting the prescribed rates, dragging tourists away from their neighbors by offering cheaper rates. As a consequence, the committee was not active at the time of the field research. In the same token, a "community development committee" aims at raising funds from tourism for social activities, but was dysfunctional at the time of research. A "Sustainable Tourism Development Committee"—initiated by a development project to provide loans for tourism-related investments—is the youngest tourism association in Thulo Syabru. Other social organizations in the village include mothers' groups and a youth club.

The people of Thulo Syabru also maintain various *informal institutions*. Households invest in reciprocal relationships, contributing time, money and food on social occasions such as funerals, births or religious festivals. However, intra-community conflicts became apparent during the group discussions. Next to a group of affluent, "well-respected" land- and livestock-owning farmers, some of which also benefit from tourism by selling farm products to the lodges, lodge owners have become a wealthy and influential group in the village. In contrast, many of the poorer farm households in the village have not yet benefitted from tourism economically. As the group discussions revealed, social change and intra-community conflicts were not only related to tourism-related inequality, but also had to do with development projects, resource use restrictions imposed by the national park, party politics and general tendencies in Nepalese society.

Although villagers claimed to be weak in vertical, *extra-community relations* during the group discussions, the survey data revealed that 53 % of households are multi-locational, i.e. have members living outside the village. Multi-locality is not necessarily a reliable indicator of bridging social capital; in Rasuwa district and other remote rural areas of Nepal, it is commonly linked to the lack of education facilities and temporary absence of children for schooling purposes. Some 75 children, both from tourism and non-tourism households have been financially supported by foreign tourists, confirming the close connection between bridging social capital—in this case friendships with foreigners—and human capital. More than half of the households have family members in Kathmandu and 10 % in a foreign country. Some lodge owners maintain contacts with foreign tour operators to sustain their tourism business.

Summing up thus far, inequality, diverging interests and declining trust within the community appear as causes of dysfunctional institutions and intra-community conflicts. At the same time, community members have managed to expand extra-community networks even beyond national boundaries. It is impossible to attribute observed changes exclusively to tourism. Without doubt, however, tourism has been an important agent of change, transforming a farm-dependent village economy into a more diversified one with new income-generating opportunities for local residents.

13.6.2 Gatlang, a Non-tourism Community in Rasuwa District

Gatlang is located just outside Langtang National Park, at an altitude of 2238 m. Like Thulo Syabru, the village is almost exclusively inhabited by ethnic Tamang. Without exception, the compact settlement consists of traditional, two-storied stone houses with a wooden roof and carved windows. As in Thulo Syabru, the economic mainstay of most households in Gatlang is mixed farming. As no household is able to produce enough food for its own consumption, animal husbandry is another important income source. Due to a lack of economic diversification and non-farm income opportunities, combined with unfavorable climatic conditions, households suffer from seasonal food and income shortages, particularly during the winter months.

Despite the scenic location of the village and its authentic Tamang character, tourism has not yet become an important sector of the village economy. Apart from one private lodge and a community lodge, no facilities and services for tourists are available in Gatlang, despite concerted efforts of government and development projects to promote tourism along the "Tamang Heritage Trail," a newly-developed trekking route. Notwithstanding the little benefit so far, people expect positive impacts of tourism and are happy about the on-going efforts to promote their village as a tourism destination.

The overwhelming majority (86 %) of households in the Gatlang sample are members of at least one *formal institution*. Most of the households are involved in community-based savings and credit groups, which have been promoted by a national NGO. As in Thulo Syabru, a "Sustainable Tourism Development Committee" was recently formed by a development project and provides tourism-related loans. In addition, there are a community forest user group, a handicraft producers' association, a cultural group, a farmers' cooperative and two youth clubs.

Solidarity, trust and social cohesion in Gatlang are promoted by a number of *informal institutions*. Conflicts, which according to the participants of the group discussions seldom occur, are settled by elected village chiefs. Institutions of mutual help are not only essential to deal with shocks such as the death of a family member or natural hazards; community members also help each other during the harvest season and with other farm-related matters through labor exchange systems. Social cohesion is strengthened by Buddhist festivals and family events such as weddings and funerals, which are jointly celebrated by the whole community. Marriage between cousins is common among Tamang families, contributing to strong kinship ties within the community.

With regard to *bridging social capital*, few households in Gatlang have family members or close relatives in Kathmandu or other urban areas of Nepal. Only 29 % of households in the sample are multi-locational, and the absence of household members is often linked to education. International labor migration as an income-generating activity is a fairly recent trend in Gatlang, as the required upfront payment to a manpower agency forms an effective barrier for most households. In the absence of vertical social capital or formal insurance, the abundance of bonding social capital in the form of horizontal associations, trust and shared norms among the community appears as a necessity.

13.6.3 Tourism and Social Capital in the Nepalese Mountains: Analysis and Discussion

Table 13.6 compares the main findings of the two mountain case studies. The average household income in the tourism village (Thulo Syabru) is twice as much as in Gatlang. In terms of relative income share, tourism is clearly more important in the local economy of Thulo Syabru (28 %) as in Gatlang (11 %). As regards *bonding social capital*, both the number of organizations and the share of households who are members in these organizations is greater in Gatlang than in Thulo Syabru. In contrast, *bridging social capital*, exemplified here by the share of multi-locational households and extra-community networks, is greater in Thulo Syabru than in Gatlang. Intuitively, the quantitative findings from the village sub-samples suggest that bonding social capital has declined and bridging social capital increased in Thulo Syabru *because of* tourism. However, these results are not significant statistically and should thus not be overvalued. To gain a deeper understanding of the causalities between tourism and social capital, the group discussions and observations in the case study villages deliver important additional information.

In Thulo Syabru, several community organizations were dysfunctional at the time of the field research. Intra-community conflicts and the decline of active village

Table 13.6 Social capital indicators for the case study communities in the Nepalese mountains

	Thulo Syabru (tourism; n = 51)	Gatlang (non-tourism; n = 70)
Mean annual cash income per household[a] (2005/2006)	US$691	US$330
Mean income share of tourism[a]	US$263 (28 %)	US$39 (11 %)
Bonding social capital		
No. of formal institutions	12	17
Share of households with membership in at least one formal institution	71 %	86 %
Bridging social capital		
Share of multi-locational households	53 %	29 %
Share of households with family/relatives in national capital	55 %	3 %

Data source: own survey

[a]Exchange rate at time of survey (November 2006); values not adjusted for purchasing power

associations have negatively affected the community's ability to act collectively, as exemplified by the failure of the lodge management committee to enforce standardized meal and accommodation rates. The latter has negative repercussions for the local tourism enterprises themselves. However, as outlined in the preceding paragraphs, the decline of bonding social capital cannot be exclusively linked to tourism; the villagers also quoted party politics and interventions of international aid organizations as reasons for intra-community conflicts and the lack of sustainability of local associations. Without doubt, an increase in income inequality among villagers can be attributed to tourism. As explained by older informants, local residents used to be more or less equal in economic terms before the advent of tourism, i.e. until the 1970s. Thereafter, tourism promoted the emergence of a new "class" of wealthy households. It seems, however, that it is not so much tourism (or the resulting income inequality) per se but rather the *decline* of tourism due to political instability in Nepal since the end of the 1990s which has aggravated local conflicts. At times when tourism thrived well in Thulo Syabru, both tourism and non-tourism households enjoyed the ample benefits of tourism, including better linkages with the "outside world" and better educational opportunities for their children. Only when tourism income started to decrease, the decline of social cohesion surfaced negatively. Fearing neighbors' competition, tourism entrepreneurs started to sabotage the rules of the lodge management committee in an effort to maximize individual returns on their sometimes considerable investments.

As compared to Thulo Syabru, life has hardly changed for the residents of Gatlang since the 1970s. Arguably, the greatest socio-economic change in the village has been brought about by the construction of a dirt road in 1989 for access to a remotely located mine, which has created a modest opportunity for local farmers to sell their crops. The lack of non-farm income opportunities, combined with unfavorable climatic conditions for farming, result in seasonal food and income shortages for almost all households. In this situation, which is typical for many remote mountain villages of Nepal, the abundant stock of social capital in the form of self-help mechanisms and other social institutions is no end in itself, but vital for survival.

Notwithstanding, a group of young, educated men in Gatlang voiced their dissatisfaction with traditional institutions and social norms (e.g. dominance of males and elderly people), regarding them as stumbling blocks to innovation and economic progress in their village. Frustrated about their limited economic prospects in the village, they were looking for ways to engage in labor migration. It is in the realm of speculation whether tourism development in Gatlang and elsewhere would be able to stop the rural exodus, which is happening in many parts of the developing world. It appears likely, however, that tourism could promote socioeconomic change in Gatlang in a similar manner as in Thulo Syabru—with comparable consequences for social capital. Once households get wealthier, the importance of community organizations may decline.

13.7 Tourism and Social Capital in the Nepalese Lowlands

13.7.1 Sauraha, a Tourism Community in Chitwan District

Sauraha is located on the shore of the Rapti River inside the buffer zone of Chitwan National Park, at an altitude of 250 m a.s.l. Originally settled by indigenous Tharu, Sauraha today is an ethnically mixed community. A small market center with modern, cemented buildings has developed along its main road. Most tourism-related businesses such as lodges, souvenir shops and restaurants are also found there. The thatched, mud-plastered farm houses on the village periphery are in stark contrast to the almost urban appearance of the touristic centre. Farming, tourism and trade are the main pillars of the local economy. Tourism is the most important economic activity in Sauraha, engaging 54 % of all households (Bachhyauli VDC 2006). In a typical (i.e. median) tourism household, tourism contributes half of total cash income. Farming is the second most important activity, involving 50 % of households in Sauraha but contributing only 9 % on average to households' cash income. Trade- and service-related activities, many of which are linked to tourism, are additional sources of household income. The importance of non-farm income among Sauraha's households reflects a larger economic diversification as compared to the mountain villages.

Tourism in Sauraha started in the mid-1970s, soon after the establishment of Chitwan National Park. The opening of a national park office with elephant stables, combined with a relatively good accessibility favored Sauraha's development as a tourist destination. Popular tourist activities include elephant rides, canoe tours, game drives, bird watching excursions, nature walks and village tours. Many hotels offer Tharu cultural shows with music and dance in the evenings. Apart from accommodation establishments, a large number of restaurants, pubs, retail shops, bicycle rentals and travel agencies in Sauraha cater to the needs of foreign tourists, as do hairdressing and massage salons, pharmacies, laundry services, internet cafes, beauty parlors, tailors and money changers, all offering employment opportunities for local residents. Most tourist facilities in Sauraha have remained family businesses. Since the end of the 1980s, more and more "outsiders," often business people from Kathmandu, bought land from locals and invested in lodges and other tourism-related businesses. Tourist arrivals at Sauraha had grown steadily to a record high of 106,000 in 2001 but sharply declined to about 67,000 visitors in 2006, again due to the political crisis in Nepal and other external factors (cf. Shakya 2009). Since then, domestic tourism has steadily grown, and Nepalese tourists now make up at least 60 % of visitors. Informants noted a shortage of skilled labor as a result of tourism decline. According to informants, many trained nature guides and hotel employees left Nepal to seek foreign employment due to the political and economic crisis. People unanimously felt that tourism had made their livelihoods more secure in many ways, despite the recent decline of tourist arrivals. However, they also admitted that tourism had unwanted effects on the local society and economy, including agricultural labor shortages and exploding land prices (cf.

Kunwar 2002). Better education, especially for girls, higher awareness with regard to hygiene and conservation, and charitable activities of foreigners, who sponsored some schools and orphanages in surrounding villages, were quoted as some of the non-material benefits of tourism. Drug abuse among youth and erosion of the indigenous culture (e.g. tourists' influence on local clothing style) were considered as negative.

An overwhelming majority (87 %) of households in the Sauraha sample are members of at least one of more than 20 *formal institutions*. According to villagers, flood prevention and control, tourism and natural resource management are aspects of life that require collective action. More than 60 % of households participate in a savings or microfinance program. There are also a number of tourism-related organizations in Sauraha, including associations of tourism entrepreneurs, tour guide associations and elephant booking offices. Furthermore, a number of charitable, political, cultural and nature conservation organizations exist in Sauraha.

As compared to the Rasuwa case studies, *informal institutions* play a less important role in Sauraha. Trust and community cohesion appear weaker than in the mountain villages, arguably due to the divergence of interests between tourism and non-tourism households. Participants in the group discussions claimed that mutual help, decision-making capacity and the economic role of women had changed positively in the past decade but did not link this to tourism. However, people's self-confidence and ability to resolve conflicts was attributed mainly, but not exclusively to tourism. Tourism was perceived as having a positive influence on the preservation of indigenous cultures and traditions. On the other hand, "individualism," "money-mindedness" and "lack of unity" were mentioned by residents as negative impacts of tourism on social cohesion.

Although only 31 % of households in Sauraha are multi-locational, the community has abundant *bridging social capital* in the form of extra-community networks. About half of respondents have family members in Kathmandu or other urban areas of Nepal. Almost a quarter (21 %) of surveyed households has relatives in a foreign country. In the group discussions, marriage to a foreigner was mentioned as a popular strategy to promote socio-economic advancement (cf. Kunwar 2002). This may explain why 41 % of international migrants from Sauraha reside in Europe (Bachchyauli VDC 2006). Overall, the results of the Sauraha case study with regard to social capital are similar to those of Thulo Syabru, indicating a large number of formal institutions and extra-community networks, and a decline of informal institutions and social cohesion.

13.7.2 Shaktikhor, a Non-tourism Community in Chitwan District

Shaktikhor is situated at an altitude of 355 m a.s.l. at the foot of the Mahabharat range. The indigenous Chepang are one of the largest ethnic groups in this culturally mixed community. The majority of dwellings lines up to both sides of an unpaved district road, which connects Shaktikhor with Nepal's main lowland highway. As in Sauraha, a central market area has developed along the road, in stark contrast to the surrounding agricultural landscape with its dispersed homesteads.

Mixed farming is the most important economic activity in Shaktikhor, involving 95 % of sample households. In terms of cash income, households depend mainly on non-farm activities such as services and trade, which on average contributed 70 % to annual income (excluding tourism). Farming contributes only 27 % of household income, thus being less important as a source of cash than in Gatlang. Corresponding income ranges suggest that inequality in Shaktikhor is not as extreme as in Sauraha.

Like in Gatlang, tourism does not yet play a significant role in Shaktikhor, despite recent efforts to promote the "Chitwan Chepang Hill Trail" as a new tourist product. Only 11 households reported to have earned some money from tourism. Tourist amenities in Shaktikhor include one lodge, home stay accommodation in eight households, a multiple-use visitor centre with a small "Chepang Museum," guiding and portering services. Apart from its good accessibility, proximity to Chitwan National Park and favorable location as the gateway to the Chitwan Chepang Hill Trail, Shaktikhor has a pleasant climate, a culturally diverse and welcoming community and is a good birdwatching destination. Against the scenic background of the Mahabharat hills, visitors can discover the rural life-style at Shaktikhor during a home stay.

Like in Gatlang, people have high expectations from tourism, despite little tangible benefit so far. They are not worried about possible negative impacts and believe that poor and wealthy households alike had a chance to benefit from tourism. Apart from economic opportunities, people expect also non-material advantages from tourism for their village.

As almost all households are involved in one of five community-based organizations, it is not surprising that 95 % of sample households quoted membership in a *formal institution*. The main purpose of these organizations is the provision of microcredit. In total, there are 17 non-governmental organizations in Shaktikhor, concerned with financial services, natural resource management, agriculture, social issues and tourism, the latter including the newly-formed "sustainable tourism development committee," a tour guide association and cultural groups.

Despite the ethnically heterogeneous population, many *informal institutions* exist in Shaktikhor. They fulfill similar functions than their counterparts in Gatlang, promoting mutual help among community members. Informants claimed that there was little conflict in the community, and *social cohesion* was reported to have improved in all dimensions.

The people of Shaktikhor are also actively involved in *extra-community networks*: Many (39 %) households are multi-locational. An equal share stated to have friends or relatives in Kathmandu and 29 % in a foreign country. The latter is the highest percentage among all case study communities.

In conclusion, the abundance of bonding social capital in Shaktikhor principally resembles the scenario in Gatlang, the non-tourism village in the hills. Unlike

Gatlang residents, however, households in Shaktikhor have benefited from better connectivity and greater diversity of the rural economy. They have thus been able to economically benefit from social capital by investing micro-loans in a range of alternative economic activities.

13.7.3 Tourism and Social Capital in the Nepalese Lowlands: Analysis and Discussion

While the absolute and relative income share of tourism is clearly larger in the "tourism village" Sauraha as compared to Shaktikhor, the number of village-based organizations and the share of households who are members of such organizations does not differ much between the two lowland villages. In absolute terms, more associations operate in Sauraha than in Shaktikhor. However, unlike in Sauraha, almost all households (95 %) in Shaktikhor are members of such organizations.

Again, the group discussions provide explanations for the strength of *bonding social capital* in Shaktikhor, and its decline in Sauraha, roughly confirming the findings from the mountains. With regard to *bridging social capital*, however, the results from the two lowland villages differ from the mountain case studies. While Thulo Syabru (mountains) appears particularly rich in extra-community relations, the non-tourism village Gatlang has a low stock of bridging social capital (cf. Table 13.6). In contrast, bridging social capital differs less between the two lowland villages in quantitative terms (Table 13.7). In qualitative terms, however, Sauraha residents tend to be better connected with other urban areas of Nepal and have more international links due to migration. Especially the latter is clearly linked to tourism.

Table 13.7 Social capital indicators for the case study communities in the Nepalese lowlands

	Sauraha (tourism; n = 77)	Shaktikhor (non-tourism; n = 61)
Mean annual cash income of households[a] (2005/2006)	US$1922	US$650
Mean income share of tourism[a]	US$1159 (60 %)	US$19 (3 %)
Bonding social capital		
No. of formal institutions	>20	17
Share of households with membership in at least one formal institution	87 %	95 %
Bridging social capital		
Share of multi-locational households	31 %	39 %
Share of households with family/relatives in national capital	20 %	39 %

Data source: own survey
[a]Exchange rate at time of survey (November 2006); values not adjusted for purchasing power

Underlined by the quantitative evidence as well as by the case study results, geographical location and related differences such as social and physical infrastructure play a major role in shaping the socio-economic impacts of tourism. In comparison with the mountain communities, livelihoods in Sauraha and in Shaktikhor are more diversified, with the major share of cash income originating from non-farm activities. Tourism clearly dominates as the most important non-farm activity in Sauraha, while the economic portfolio in Shaktikhor is broader and has retained closer links with the traditional farm economy.

The differences in bonding and bridging social capital between the mountains and the lowlands can also be explained by geographical location. For instance, establishment and operation of formal associations is generally easier in the lowlands, due to shorter distances, better communication means and lower transport costs.

13.8 Conclusions and Lessons Learnt

This chapter proposed the concept of social capital as a framework to study social consequences of tourism, based on case study evidence from poor, rural communities of Nepal. We hypothesized that tourism is able to alter different dimensions of social capital, such as formal village associations, social cohesion and extra-community relations. The empirical study was conducted among four rural communities in two different geographical regions of Nepal. In a first step, aggregate household data from the case study communities were analyzed statistically, revealing no systematic impact of tourism on households' social capital but rather suggesting an effect of contextual variables on social capital.

In a second step, the spatial scope of the study was narrowed down, analyzing the evidence from the villages separately and comparing tourism and non-tourism as well as mountain and lowland communities. Findings from the mountains indicate a decline of bonding social capital and a considerable increase in bridging social capital in the tourism village, and some of these observations were linked to tourism. In contrast, bonding social capital plays an important economic role in the non-tourism community as informal insurance. The research results demonstrate the massive socio-economic transformation which can be brought about by tourism in remote village societies. In contrast, social capital has remained a vital asset in the agriculture-dependent mountain community, where tourism does not yet play an important role.

The case studies from the Nepalese lowlands generally confirm the findings with regard to bonding social capital, but did not identify any quantitative difference in bridging social capital between tourism and non-tourism villages in the lowlands. Complementary evidence suggests, however, that tourism has an impact on the quality of bridging social capital in the lowlands, e.g. on international networks.

The research results support our theoretical assumptions with regard to tourism's impact on social capital. Tourism has promoted the formation of new institutions and created opportunities for participation in vertical, extra-community networks. However, tourism can also exacerbate local conflicts and impair the functioning of indigenous institutions. Such erosion of social capital can have negative economic consequences, if it reduces the ability of communities to act collectively. The study also shows that bonding social capital can be substituted by bridging social capital, and both forms of social capital can be substituted by other assets. Finally, the study highlights the importance of contextuality for any empirical study on tourism's impacts. More generally, as shown in this chapter, the (potential) impacts of an economic activity or policy intervention are inseparable from and moderated by the geographical setting and other contextual variables.

Several lessons for sustainable tourism development emerge from this chapter. First, the concept of social capital is an adequate framework to analyze the social and cultural impacts of tourism on destination communities. Several types of social capital can be distinguished, with different socioeconomic implications. Tourism is able to alter the social capital of destination communities; it may enhance one type of social capital while eroding another. This was exemplified by the transformation from horizontal (bonding) to vertical (bridging) social capital in the tourism villages. Linked to its economic effects, tourism is able to also change other types of assets, e.g. human capital (through improved education) and financial capital (through accumulation of savings), thereby reducing the importance of social capital.

Pre-existing social capital in the form of local institutions and social cohesion facilitates bottom-up, participatory tourism planning and communities' commitment to tourism development. It can also be a tourist attraction in its own right, as illustrated by Buddhist rituals and colorful festivals that are observed collectively throughout Nepal. However, policy-makers and tourism planners must carefully analyze and respect the values that various social groups in destination communities attach to social capital. As communities advance economically, social change is inevitable and not just linked to tourism. Normative judgments whether such changes are "good" or "bad" have to be left to the concerned communities, however. It is up to them to opt for or against tourism as a development path and accept the social change that not only comes along with tourism but with any economic transformation.

References

Ashley, C., and S. Maxwell. 2001. Rethinking rural development. *Development Policy Review* 19(4): 395–425.

Bachhyauli VDC (Village Development Committee). 2006. *Population survey, ward 2, Shrawan 2063 (July/August 2006)*. Bachhyauli.

Bachmann, P. 1988. *Tourism in Kenya. Basic need for whom?* Frankfurt am Main/New York/Paris: Berne.

Beugelsdijk, S., and S. Smulders. 2003. *Bridging and bonding social capital: Which type is good for economic growth?* European Regional Science Association (ERSA) conference papers (ersa03p517). http://www.eea-esem.com/papers/eea-esem/2003/119/EEA2003.PDF

Boo, E. 1990. *Ecotourism: the potentials and pitfalls*, vol. 1. Washington, DC: World Wildlife Fund.

Bourdieu, P. 1986. The forms of capital. In *Handbook of theory and research for the sociology of education*, ed. J.G. Richardson, 241–258. New York: Greenwood Press.

CBS (Central Bureau of Statistics), National Planning Commission Secretariat and His Majesty's Government of Nepal. 2005. *Poverty trends in Nepal (1995–96 and 2003–04)*. Kathmandu: Central Bureau of Statistics.

Claiborne, P. 2010. *Community participation in tourism development and the value of social capital. The case of Bastimentos, Bocas del Toro, Panamá*. Master thesis. Gothenburg: Graduate School of Tourism and Hospitality Management.

Cohen, E. 1988. Authenticity and commoditization in tourism. *Annals of Tourism Research* 15: 371–386.

Coleman, J. 1988. Social capital in the creation of human capital. *American Journal of Sociology* 94(Suppl): 95–120.

Coleman, J. 1990. *Foundations of social theory*. Cambridge, MA: Harvard University Press.

Coppock, R. 1978. The influence of Himalayan tourism on Sherpa culture and habitat. *Zeitschrift für Kulturaustausch* 28(3): 61–68.

De Kadt, E. (ed.). 1979. *Tourism: Passport to development? Perspectives on the social and cultural effects of tourism in developing countries*. New York: Oxford University Press.

Esman, M., and N. Uphoff. 1984. *Local organizations: Intermediaries in rural development*. Ithaca: Cornell University Press.

Feldman, T.R., and S. Assaf. 1999. *Social capital: Conceptual frameworks and empirical evidence. An annotated bibliography*. Social capital initiative working paper 5, Washington, DC.

Graburn, N., and J. Jafari. 1991. Introduction: Tourism social science. *Annals of Tourism Research* 18: 1–11.

Granovetter, M. 1985. Economic action and social structure: The problem of embeddedness. *American Journal of Sociology* 91: 481–510.

Grootaert, C. 1998. *Social capital: The missing link?* Social capital initiative working paper 3, Washington, DC.

Grootaert, C., and T. Van Bastelaar. 2001. *Understanding and measuring social capital: A synthesis of findings and recommendations from the social capital initiative*. Social capital initiative working paper 24, Washington, DC.

Gurung, D. 1995. *Tourism and gender. Impact and implications of tourism on Nepalese women*. MEI (Mountain Enterprises and Infrastructure) discussion paper series 95/3, Kathmandu.

Harrison, D. (ed.). 1992. *Tourism and the less developed countries*. London: Belhaven Press.

Hashimoto, A. 2002. Tourism and sociocultural development issues. In *Tourism and development. Concepts and issues*, ed. R. Sharpley and D.J. Telfer, 202–230. Clevedon: Channel View Publications.

Hirschman, A. 1984. *Getting ahead collectively: Grassroots organizations in Latin America*. New York: Pergamon Press.

Kunwar, R.R. 2002. *Anthropology of tourism. A case study of Chitwan-Sauraha, Nepal*. Delhi: Adroit.

Macbeth, J., D. Carson, and J. Northcote. 2004. Social capital, tourism and regional development: SPCC as a basis for innovation and sustainability. *Current Issues in Tourism* 7(6): 502–522.

McGehee, N.G., et al. 2010. Tourism-related social capital and its relationship with other forms of capital: An exploratory study. *Journal of Travel Research* 49(4): 486–500.

MCTCA, UNDP and TRPAP (Tourism for Rural Poverty Alleviation Programme). 2005. *Tourism resource mapping profile. Rasuwa district*, Kathmandu.

MCTCA, UNDP and TRPAP. 2006. *Tourism resource mapping profile. Chitwan district*, Kathmandu.

Mihalič, T. 2002. Tourism and economic development issues. In *Tourism and development. Concepts and issues*, ed. R. Sharpley and D.J. Telfer, 81–111. Clevedon: Channel View Publications.

Mitchell, J., and C. Ashley. 2007. *Can tourism offer pro-poor pathways to prosperity?* ODI briefing paper 22, June 2007, London.

Mitchell, J., and C. Ashley. 2010. *Tourism and poverty alleviation. Pathways to prosperity*. London: Earthscan.

Narayan, D., and L. Pritchett. 1997. *Cents and sociability. Household income and social capital in rural Tanzania*. World Bank policy research working paper 1796, Washington, DC.

Nepal, S.K. 2005. Tourism and remote mountain settlements: Spatial and temporal development of tourist infrastructure in the Mt Everest Region, Nepal. *Tourism Geographies* 7(2): 205–227.

North, D.C. 1990. *Institutions, institutional change, and economic performance*. New York: Cambridge University Press.

NTB and TRPAP (Nepal Tourism Board and Tourism for Rural Poverty Alleviation Programme). 2003. *Household socio-economic survey in TRPAP areas*. Submitted by Prabidhik Paramarshak Toli Pvt. Ltd. to NTB for TRPAP, Kathmandu.

Olson, M. 1982. *The rise and decline of nations: Economic growth, stagflation, and social rigidities*. New Haven: Yale University Press.

Ostrom, E. 1990. *Governing the commons. The evolution of institutions for collective action*. Melbourne: Cambridge University Press.

Ostrom, E. 2000. Social capital: A fad or a fundamental concept? In *Social capital: A multifaceted perspective*, ed. P. Dasgupta and I. Seragilden, 172–214. Washington, DC: World Bank.

Portes, A. 1998. Social capital: Its origins and applications in modern sociology. *Annual Review of Sociology* 24: 1–24.

Putnam, R.B. 1993. The prosperous community—Social capital and public life. *American Prospect* 13: 35–42.

Putnam, R.B. 1995. Bowling alone: America's declining social capital. *Journal of Democracy* 6(1): 65–78.

Scoones, I. 1998. Sustainable rural livelihoods. A framework for analysis. IDS working paper 72, Brighton.

Shakya, M. 2009. *Risk, vulnerability and tourism in developing countries: The case of Nepal*, Bochum studies in international development, vol. 56. Berlin: Logos Verlag.

Southgate, C., and R. Sharpley. 2002. Tourism, development and the environment. In *Tourism and development. Concepts and issues*, ed. R. Sharpley and D.J. Telfer, 231–262. Clevedon: Channel View Publications.

Swift, J. 1989. Why are rural people vulnerable to famine? *IDS Bulletin* 20(2): 8–15.

Telfer, D.J. 2002. Tourism and regional development issues. In *Tourism and development. Concepts and issues*, ed. R. Sharpley and D.J. Telfer, 112–148. Clevedon: Channel View Publications.

Telfer, D.J., and R. Sharpley. 2008. *Tourism and development in the developing world*. New York: Milton Park.

Turner, L., and J. Ash. 1975. *The golden hordes. International tourism and the pleasure periphery*. London: Constable.

UNWTO (UN World Tourism Organization). 2002. *Tourism and poverty alleviation*. Madrid: UNWTO.

Uphoff, N. 1993. Grassroots organizations and NGOs in rural development: Opportunities with diminishing states and expanding markets. *World Development* 21(4): 607–622.

Uphoff, N. 2000. Understanding social capital: Learning from the analysis and experience of participation. In *Social capital: A multifaceted perspective*, ed. Partha Dasgupta and Ismail Serageldin. Washington, DC: World Bank.

Vorlaufer, K. 1996. *Tourismus in Entwicklungsländern. Möglichkeiten und Grenzen einer nachhaltigen Entwicklung durch Fremdenverkehr*. Darmstadt: Wissenschaftliche Buchgesellschaft.

Wall, G., and A. Mathieson. 2006. *Tourism: Change, impacts and opportunities*. Harlow: Pearson Prentice Hall.

Wells, M., K. Brandon, and L. Hannah. 1992. *People and parks. Linking protected area management with local communities*. Washington, DC: World Bank.

Wiggins, S., and S. Proctor. 2001. How special are rural areas? The economic implications of location for rural development. *Development Policy Review* 19(4): 427–436.

Woolcock, M. 1998. Social capital and economic development: Toward a theoretical synthesis and policy framework. *Theory and Society* 27: 151–208.

Woolcock, M., and D. Narayan. 2000. Social capital: Implications for development theory, research and policy. *The World Bank Research Observer* 15(2): 225–249.

World Bank. 2006. *Nepal: Resilience amidst conflict. An assessment of poverty in Nepal, 1995–96 and 2003–2004*. Washington, DC: World Bank.

WTTC (World Travel & Tourism Council). 2007. *Nepal Travel & Tourism. Navigating the path ahead. The 2007 Travel & Tourism Economic Research*. London: WTTC.

WTTC. 2014. *Travel & Tourism Economic Impact 2014. Nepal*. London: WTTC.

Part IV
Conclusion

Chapter 14
The Way Forward

Keith Bosak and Stephen F. McCool

Abstract This concluding chapter begins by revisiting the Case of the Nanda Devi Biosphere Reserve (NDBR) and its inhabitants from Chap. 3; this time highlighting the challenges involved in implementing sustainable tourism. Many of the same challenges are being faced by Whitefish, Montana and other places that either attract tourists or hope to do so. Next, the chapter will re-visit the frameworks and concepts introduced by the contributing authors to synthesize common ideas and tools that are useful in re-framing sustainable tourism at the ground level. This 'toolkit' of frameworks and ideas will be applied to the case of the NDBR to show how a re-framing of sustainable tourism might help in addressing some of the many challenges faced in sustainable tourism development.

Keywords Sustainable tourism • Planning • Scale • Socio-ecological systems • Resilience

14.1 Introduction

It has been nearly 15 years since the communities of the NDBR gathered to develop the Ecotourism and biodiversity conservation declaration and although tourism exists in the region, there have been and continue to be major challenges. Many of these challenges stem from the drastic changes that have occurred in these remote Himalayan villages in the last 15 years. On my first visit to the NDBR, the villages were struggling with a subsistence lifestyle. Most families were farmers with the men of the family (and sometimes the women) working day labor to earn a meager income. Tourism, and particularly ecotourism as a sustainable livelihood, was a welcome opportunity to earn some money to perhaps spend on children's education, better food, clothing or other essentials. Ecotourism was also seen as a way to keep younger generations in the villages and provide them with a viable livelihood based in place, thus continuing the distinctive local (Bhotiya) culture and traditions.

K. Bosak (✉) • S.F. McCool
University of Montana, Missoula, MT, USA
e-mail: Keith.Bosak@umontana.edu; Steve.McCool@cfc.umt.edu

S.F. McCool, K. Bosak (eds.), *Reframing Sustainable Tourism*, Environmental Challenges and Solutions 2, DOI 10.1007/978-94-017-7209-9_14

In 2002, on my first visit, there was electricity but it was intermittent and unreliable. Many families actually refused to be hooked up for fear of being overcharged by the government. There were no TVs, electric appliances, propane cook stoves, cell phones, computers or telephones. Virtually no families owned a car and mobility was based on a system of share taxis that ran routes up and down the valley to Joshimath, the nearest town. Today, most families have electricity and along with it, satellite television and various electrical appliances. Cell phones are almost ubiquitous and although the signal is still spotty, people can call from most villages. The younger generations have smartphones and frequent Facebook and other social media. The prioritization of education means that families who can afford it, send their children off to the towns and cities for school. These children do not grow up in the villages and consequently do not gain the wealth of local knowledge that their parents and grandparents had. With the outmigration of the younger generation has come the in-migration of consumerism. Western clothing is becoming more common as are other consumable items such as those mentioned above. The shift to consumerism means that more and more income is necessary to keep up with that type of lifestyle. As such, many villagers have turned to commercializing their agriculture, growing potatoes and kidney beans for cash, only to buy food from the market.

There has also been a boom in the collection of *Cordyceps sinensis*, an alpine medicinal fungus that can fetch as much at $7000 US a kilogram. Although the collection season is limited to 2 months, the activity is so lucrative that many families are actually able to build new homes and some are now purchasing their own automobiles. Unfortunately, the collection of this medicinal fungus is not being undertaken sustainably as was common for such resources in the past. The rush to collect *Cordyceps sinensis* is a disturbing and recent example of overexploitation of the environment. Meadows where the fungus is collected are quickly being degraded by overuse and the fungus itself is being overharvested. What does all of this have to do with sustainable tourism?

What the case of the NDBR highlights is that the spread of global capitalism is vast and ever increasing with tourism being only one driver of change. In the midst of these changes, the communities of the NDBR have been developing their own form of place-based ecotourism, dependent on the vast natural resources that they themselves protected for hundreds of years and that the government now protects through conservation efforts. Originally, the push for sustainable tourism in the form of ecotourism in NDBR was driven by a combination of political factors and the drive for social and environmental justice. Today, income is the main driver for ecotourism in the region. However, the collection of *Cordyceps* now rivals and often out-competes ecotourism as a livelihood activity because it is so lucrative (but most likely unsustainable). What we have learned from several of the preceding chapters in this book is that global capitalism and the natural resources on which tourism depends are often at odds and this produces many challenges for sustainable tourism. This leads to the question: What tools could be used to re-frame sustainable tourism on the ground?

14.2 Tools for Re-framing Sustainable Tourism

The tools for re-framing sustainable tourism introduced in this concluding chapter are drawn from ideas introduced by the chapter authors in this volume and seek to provide specific approaches to re-frame sustainable tourism at the ground level. As evidenced by the case of the NDBR, the world is not the same as it was in the 1980s when the concepts of sustainable development and sustainable tourism were first introduced. The breakup of the Soviet Union and the shift of China and India to capitalist economies ushered in a new era of globalization not seen previously. Put simply, the political, economic and social environment that sustainable tourism operates within is not the same and it will continue to change. Conceptualizing sustainable tourism as three spheres of activity (economic, social and environmental) where the goal is to minimize negative impacts of tourism in each sphere is largely ineffective on the ground and is not suited to the ever-changing and complex world in which we live today. The tools introduced here and derived from the contributors to this book are intended to address both change and complexity. These tools are operationalized through the example of NDBR in order to show how each tool can enhance the sustainability of tourism at the local scale rather than just sustaining tourism.

These tools are all based on a few foundational tenets. The first is that the defining characteristic of tourism is that it changes the places where it occurs, often to meet its own needs. And thus, tourism is as much an agent of change as it is a result. The second is that planning is necessary for any type of sustainability for sustainable tourism does not occur spontaneously; it must be planned in a way that is inclusive, thoughtful and adaptive; it must be respectful of not only a diversity of perspectives but also of underlying and widely shared values, and it must recognize the critical role that local and regional institutions frame what is possible. Planning in this sense allows for deliberation, idea generation, innovation and visioning. It also provides a clear set of goals and actions to achieve those goals. Planning helps deal with complexity when it is coupled with implementation and monitoring. If the desired outcomes are not being achieved, the plan must be changed. Adaptive planning allows for resilience in the face of change and complexity.

The remainder of the chapter will focus first on introducing tools that are designed to re-frame sustainable tourism, allowing stakeholders to view sustainable tourism through a different lens, one that seeks to dive deeper in order to get at some of the fundamental assumptions of sustainable tourism in the broader context of change and complexity.

The first step in diving deeper is viewing sustainable tourism as part of a broader socio-ecological system, one that exists at multiple scales and in multiple contexts. Each tourism destination has its own unique context based on the physical environment, culture, history, economics and politics of the location. In addition, each location is connected to other locations (some are tourists destinations and some are not) through flows of resources, people, money, technology, and ideas. In addition, tourists, locals, tourism providers, policy makers and conservationists all bring in a unique perspective and set of motivations, some of which can be conflicting. Socio-ecological systems are complex, adaptive and ever-changing. Understanding sustainable tourism in the context of socio-ecological systems helps to keep the focus on the intimate linkages between society and the environment rather than artificially separating out each component (economic, social, environmental) as is common in conventional sustainable tourism practice. Such artificial separation can and often does lead to unwanted and unanticipated feedbacks.

Understanding socio-ecological systems requires focusing on the resilience of these systems as well as on scale and scalar processes both spatial and temporal. Socio-ecological systems are managed for resilience because these systems occur in an ever-changing and unpredictable environment, with a large array of forces and processes that push the system toward unacceptable thresholds. In addition; systems are nested within other systems producing a multi-scalar array of systems that constantly interact across space and time. These scalar interactions add complexity and remove predictability and often challenge the resilience of systems at various scales. The global can affect the local and vice versa. Scale is also a representation of power. This is evidenced through various institutions operating at different scales. We talk in terms of the global economy and immediately think of the Worldbank and WTO or of conservation and think of UNESCO and IUCN. National governments wield their own power through laws and policies as do regional and local governments.

Looking at the case of the NDBR from a systems perspective with a focus on resilience and scale gives us an idea of how sustainable tourism might be developed to enhance the stability of the system at the local level while having sensitivity to feedback occurring at other scales. The rugged and dynamic environment of the Himalaya is a main feature of this socio-ecological system. The Garhwal region where NDBR is located has vast altitudinal gradients that allow for high climatic range and consequentially biological diversity. The Himalayan Mountains affect the weather and gather and store critical amounts of freshwater that then runs out to the Bay of Bengal through the Gangetic plains where hundreds of millions of people depend on its flow for their survival. Within this context, the local people living in what is now the buffer zone of the NDBR have historically depended on taking advantage of the high biodiversity to sustain themselves. This included grazing of sheep and goats, subsistence agriculture, gathering of plants, hunting, and trading between the Tibetan Plateau and the Gangetic Plains. It also involved seasonal migration for most villages.

Over time, the culture of the Bhotiya people developed out of their interactions with the place in which they live. It is this intimate connection between place and culture that the systems approach is ideal for understanding. Today's system still relies on the environment as its centerpiece, however, the interactions look very different. The environment is protected by national and international biodiversity conservation policies. These policies have severely restricted local access to natural resources thereby making sustainable livelihoods difficult if not impossible. In the villages of the NDBR, people are increasingly entering the global economy, often with money earned from the sale of commercial crops, collection of the alpine medicinal fungus, *Cordyceps sinensis*, and secondarily from tourism wages and day labor.

There is great concern for the changes taking places among villagers as older generations watch local knowledge disappear and along with it, their culture. The younger generations would like to be able to live in their villages but there are no stable economic opportunities. Sustainable tourism in this context can provide a linkage between the environment and local culture that could enhance resilience of social-ecological systems. Through sustainable tourism, in this case ecotourism, the villages of the NDBR seek to maintain their culture and as such, their access to the resources on which their culture depends. In this case, the reasons for pursuing sustainable tourism go far beyond economic development and environmental conservation.

Re-framing sustainable tourism also necessitates taking a broader approach to sustainability; one that focuses on well-being and quality of life as major goals. As evidenced in the case of the NDBR, well-being and quality of life are not necessarily related to economic development and as human societies run up against the limits of what the environment can supply for the ever-increasing appetite of global capitalism, there must be other metrics with which to measure development (and sustainability). In the case of the NDBR, well-being and quality of life are related to the resilience of the socio-ecological system. This can be seen in the push to develop ecotourism that seeks to conserve bio-cultural diversity. As such, the people of the NDBR are trying to keep their culture in tact by maintaining their close and reciprocal relationship with their local environment. At the same time, social bonds formed through various relationships within and between families, clan groups and villages remain important.

Another factor that contributes to resilience in social-ecological systems is human and social capital, the relationships and connections we hold with others. Significantly, as the world changes, so does the need for different forms of human and social capital in the pursuit of resilience. While tourism has funded, through increased income, higher levels of educational attainment, it is possible to look at tourism more specifically as a means to enhance human and social capital. We have learned that in a globalized world, connections and networks are important, and tourism can be viewed as a tool to enhance those connections and networks. Tourists can be seen as clients, but they can also serve as means to expand networks and build relationships in the larger civil society.

While the drive to develop ecotourism by the people of the villages of the NDBR might seem on the surface to be a push for economic development that also serves to protect important natural resources, it is really an effort to maintain a healthy socio-ecological system, one that supports social structures and relations with the environment that ultimately produces a rich and unique culture, rooted in place. At the same time, people are struggling with the increasing influence of global capitalism and along with it, the conversion to consumerism. The draw to a consumer economy is driven by media and promises an 'easy' life. Younger generations, educated in the cities are easily influenced by the opportunity to accumulate consumer goods and material wealth. What we are seeing in this case is a transformation whereby the social system is becoming subsumed within the economic system, a dangerous shift that ultimately leads to environmental degradation and the separation of humans from nature. In this case, a re-framing of sustainable tourism can assist in understanding this shift and help communities deliberate on their trajectory of development going forward.

Below is a set of questions to help communities think about sustainable tourism.

- Why is the community pursuing sustainable tourism?
- What is to be sustained and why, and what is the ultimate goal?
- What do communities want from sustainable tourism?
- What will be the strategy for sustainable tourism development at the community-level?

These questions need to be answered with thoughtful deliberation. While on the surface, it might seem like communities are pursuing sustainable tourism for economic development, there is a need to dive deeper and ask questions such as: Economic development for what purpose? Many times economic development is sought in order to enhance quality of life (particularly in consumer-driven cultures). However, if we prioritize social relations over economic relations, quality of life looks different and is often tied to healthy relationships and social networks that not only build social capital but also enhance culture. This is the case for the villages of the NDBR. Ultimately (as evidenced through the Ecotourism and Biodiversity Conservation Declaration) the villages of the NDBR are pursuing sustainable tourism to preserve and enhance their culture. To do this, requires maintaining a healthy and mutually beneficial relationship with the local environment. This leads to the next question.

The second question communities can ask themselves is: What is to be sustained and why? Is it tourism that is to be sustained? Economic development? The environment? A way of life? Again, these are important questions for deliberation. While on the surface, it might seem like it would be beneficial to sustain economic development, ultimately, that produces environmental degradation either at the local scale or at other scales both temporal and spatial. Diving deeper through deliberation on this question allows for the question of why that particular thing is being sustained. In the case of the NDBR, I argue that what people are trying to sustain is a resilient socio-ecological system for the purpose of maintaining their way of life and their culture. Searching for resilience is particularly significant in a globalized world where connections are not only difficult to understand, but also events 12 time zones away from a community may deeply impact it. While livelihoods might change and they have in this region many times, the key factor is maintaining the socio-ecological system (relations with nature situated in place). It is this relationship between social and ecological systems that creates, maintains and enhances culture. Sustainable tourism can be another livelihood that serves this purpose for the people of the NDBR.

This leads to the third question: What do communities want from sustainable tourism and why? Do communities want economic development in the form of jobs? Do communities want better healthcare and education? Do they want better opportunities for women and girls? This question elicits a myriad of responses and again, we can go beyond economic motivations and ask: Why? Why does the community want jobs? This is a common answer in many places trying to develop sustainable tourism. In NDBR when this question is asked, the answer is often that people want jobs. However, if one asks why, the responses often center on being able to support the family, provide for education and healthcare and to stay in the village where their culture is rooted. Even younger generations often say that they would prefer to stay and live in the village but there are no opportunities for them so they must leave. What is important to understand is that while the superficial motivation might be for job creation, asking why elicits deeper responses that are often not motivated by economic interests.

The final question is: What will be the strategy for sustainable tourism development at the community level? This question is perhaps the most difficult to answer and requires an integration of scale along with a socio-ecological systems approach. Communities must ask themselves: What is the time-scale for sustainability? How is the development of sustainable tourism in this place affected by larger scales and how does it affect other scales (larger and smaller) in terms of their sustainability? How can sustainable tourism be developed in such a way as to maintain or enhance resilience in socio-ecological systems? How much and what types of change are acceptable and over what time period? These are difficult questions and they will be answered differently in every community. Fortunately, many of the tools introduced in this book such as adaptive planning and governance, environmental supply, and guidelines for tourism operators offer ways for communities to answer this final and very important question.

14.3 Some Concluding Thoughts

The purpose of this book is to re-frame sustainable tourism to make it more relevant in today's ever changing, complex and uncertain world. A secondary purpose is to provide tools to aid in developing sustainable tourism as it has been re-framed. It seems appropriate then to conclude with a summary of suggestions for re-framing sustainable tourism.

Take a broader approach and dive deeper. The events we see are like the tip of an iceberg: there are often foundational patterns and structures that influence our ideas and actions. And so it is with the notion of sustainable tourism. Oftentimes, sustainable tourism is based in ideas of economic development. However, if we go deeper, we often find that economic development is being pursued for a whole host of non-economic reasons. This is why it is also necessary to take a wider approach to sustainability that focuses on ideas of well-being and quality of life where tourism is a tool to pursue those goals.

Utilize a systems approach to understand the dynamics of sustainable tourism. The answer to the question of what is to be sustained is all too often either tourism or development. If we view sustainable tourism through the lens of socio-ecological systems, resilience becomes a key concern. As such, the answer to the question of what is to be sustained becomes the resilience of the socio-ecological system. And if that indeed is what we seek, we must ask how can tourism contribute to that resilience?

Realize and integrate scale into sustainable tourism as a core concept for understanding system dynamics. Sustainable tourism requires a long-term perspective and the impacts of human activities are often felt great distances from where they occur. In addition, scale and power are intimately related and understanding scale and scalar configurations also highlights power relations that are central to understanding system dynamics at all scales.

Planning is still a key feature in sustainable tourism. Tourism will not be sustainable unless it is planned. The ultimate goal of planning is to change the future. Planning allows for deliberation, prioritization of values, and visioning for the future to set goals. It also helps develop strategies and actions necessary for achievement of those goals. Planning needs to be participatory and adaptive. It also needs to be coupled with proper management and monitoring so that any unwanted or un anticipated changes in the sustainable tourism system can be remedied.

So here we are: at the beginning of an infinite future, thinking of what it is that tourism should sustain, and reframing this concept so it is more relevant for the challenges and opportunities confronting humankind in the twenty-first century. By focusing on what is to be sustained, we think more explicitly about the future and about the opportunities presented to our descendants. Will their opportunities be as great as we hold before us, or will they be more limited?

CPI Antony Rowe
Chippenham, UK
2017-07-17 21:48